T0384221

Supersymmetry and Superstring Theory with Engineering Applications

Supersymmetry and Superstring Theory with Engineering Applications

Harish Parthasarathy
Professor
Electronics & Communication Engineering
Netaji Subhas Institute of Technology (NSIT)
New Delhi, Delhi-110078

CRC Press is an imprint of the
Taylor & Francis Group, an **informa** business

First published 2023
by CRC Press
4 Park Square, Milton Park, Abingdon, Oxon, OX14 4RN

and by CRC Press
6000 Broken Sound Parkway NW, Suite 300, Boca Raton, FL 33487-2742

CRC Press is an imprint of Informa UK Limited

British Library Cataloguing-in-Publication Data
A catalogue record for this book is available from the British Library

ISBN: 9781032384115 (hbk)
ISBN: 9781032384122 (pbk)
ISBN: 9781003344933 (ebk)

DOI: 10.1201/9781003344933

Typeset in Arial, Minion Pro, Times New Roman
by Manakin Press, Delhi

Preface

There are several laws of physics in a scattered form like Einstein's law of gravitation that relates the metric tensor of curved space-time to the energy-momentum tensor of matter and radiation, the Klein-Gordon equations for a massive spinless scalar field, Dirac's relativistic wave equation for the electron interacting with the vector potential of the electromagnetic field, Maxwell's equations for the electromagnetic field and more generally, the Yang-Mills non-Abelian gauge field equations coupled to the matter field equations. One can in fact at a classical level unify these field equations by including the energy-momentum tensor of the matter field (comprising the electrons, positrons described by the Dirac field and the nuclear particles like the proton and the neutron or more precisely the leptons and the quarks out of which these particles are built and described by the Yang-Mills field and the scalar massive Klein-Gordon field), the electromagnetic radiation produced by the massless photons, and the Yang-Mills gauge fields (that appear in the form of the W and Z Bosons in the electro-weak theory as propagators of the nuclear forces) on the right side of the Einstein field equations. At the quantum level, one associates an elementary particle with each of these fields and according to a fundamental law of nature, such particles appear either as Bosons having integral spin or as Fermions having half integral spin. The spin 2 graviton that communicates the gravitational field, the Klein-Gordon particle associated with the massive scalar field, the photon that communicates the electromagnetic field between charges and the non-Abelian gauge particles turn out to be Bosons while the electron and the positron that are associated with the Dirac matter field and the non-Abelian matter fields are Fermions. Hitherto, the laws of physics obtained by adding all the Lagrangians for these fields with interactions included possess three kinds of invariance/symmetry: One invariance under space-time diffeormorphisms, under local Lorentz transformations and finally gauge invariance. However, what is missing in all these symmetries is an invariance under Boson-Fermion exchange. Supersymmetry fills up this gap marvellously and thus enables us to describe nature with a maximal set of symmetries by associating a Fermionic superpartner to every Boson and conversely a Bosonic superpartner to every Fermion.

A superfield is simply a fourth degree polynomial in four anticommuting Fermionic coordinates, or more precisely Majorana Fermionic coordinates (A Majorana Fermion is a Fermionic variable whose complex conjugate is simply obtained as a certain linear transformation of itself) with each coefficient in this polynomial being a function of the four space-time Bosonic coordinates. An infinitesimal supersymmetry transformation is then simply a super-vector field, ie, a derivation in the Bosonic and Fermionic coordinates. These infinitesimal transformations taken along with the generators of the Poincare algebra (namely, space-times translations or equivalently their infinitesimal generators-the energy-momentum four vector, Lorentz boosts and the angular momenta or equivalently space-time rotations) define a super-Lie algebra so that the anticommutator of two super-vector fields becomes a linear combination of the Bosonic four momentum operators, ie, space-time derivatives. When a super vector field act on a super field, the components of the superfield undergo infinitesimal transformations and certain combinations of these components under these supersymmetry transformations change by a perfect space-time divergence which means that an action obtained by integrating this combination over space-time remains invariant. Such Lagrangians are said to be supersymmetric, ie, in addition to the three kinds of symmetries/invariances described earlier, they also are invariant under Boson-Fermion exchange.

It should be noted that some of the components of the superfield are Bosonic fields like the scalar and the electromagnetic field, the non-Abelian gauge field, and the others are Fermionic fields like the Dirac field and the gaugino field and a supersymmetric transformation of a Bosonic field is a Fermionic field and vice versa. Since supersymmetric Lagrangians now have four symmetries–Local Lorentz, Diffeormorphism, gauge and supersymmetry, they furnish us with the hope of a realistic theory that unifies all the known fields of nature. When gravitation is included, then one requires the action to have local supersymmetry which is a stronger requirement than global supersymmetry. In that case, the Bosonic graviton field appears along with its superpartner–the Fermionic spin 3/2 gravitino field and we obtain a unified picture of all the laws of nature.

A complete law of nature must have a maximal group of symmetries which means by Noether's theorem that such a law contains a maximal number of conservation laws. Supersymmetry fulfils this by demonstrating the conservation of a super-current field which incorporates charge conservation, energy momentum conservation and many others.

This book gives the reader an introduction to the vast subject of supersymmetry along with many examples having engineering applications like the design of quantum unitary gates using supersymmetric actions, Bosonic and Fermionic noise in quantum systems using the Hudson-Parthasarathy quantum stochastic calculus, superstring theory applied to the quantum mechanics of neurons and supersymmetric quantum filtering theory which can for example be used to filter out the noise in a cavity resonator electromagnetic field produced by the presence electrons and positrons in a bath surrounding it. Simplified versions of super-Yang-Mills theory with gauge and gaugino fields, both transforming under the adjoint representation of the gauge group and elementary super-gravity models have also been introduced. All through the book, emphasis is laid upon exploiting the supersymmetry existing in the nature of Boson-Fermion exchange in designing engineering systems like quantum computers and analyzing the performance of systems in the presence of supersymmetric quantum noise.

The intricacies involved in understanding the calculus of Bosonic and Fermionic variables and supervector fields as representations of the Boson-Fermion superalgebra, are clearly explained including facts like how supersymmetry predicts that the mass of the Dirac electron should depend upon the Bosonic scalar field.

The construction of supersymmetric matter and gauge field Lagrangians using the calculus of Chiral superfields (both matter and gauge) with clear explanation of how these Lagrangians generalize the already well known matter and gauge Lagrangians appearing in non-Abelian Yang-Mills theories that possess only the three kinds of symmetry but not supersymmetry is explained.

The notion of super-field equations as a unified way of writing the field equations for a supersymmetric Lagrangian has been given some space.

Analysis of superstrings in the Bosonic and Fermionic Fock space has also been given some attention. Aspects of renormalization theory have been dealt with for completeness in the presentation. Supersymmetric path integrals for drawing Feynman diagrams with the hope of describing the interaction of elementary particles with their superpartners has been explained. This may enable the engineer to design larger sized quantum gates for supercomputers in view of utilizing both the Bosonic and Fermionic degrees of freedom

Local supersymmetry has been discussed in the context of superparticles, superstrings and super-gravity. In local supersymmetry, the action is invariant even when the Majorana Fermionic parameter appearing in the definition of a supersymmetry transformation becomes a function of the space-time variables. Thus, if nature really follows local supersymmetry then our quantum gates designed based on such Lagrangians using the Feynman path integral for fields would have nice symmetry properties.

It is hoped that with this introduction, the reader will be able to understand some more important problems involving supersymmetry breaking at high energy scales and how this is communicated to lower energy scales via messenger superfields or by gravitational fields which is the reason why so far superpartners of the particles in the standard model are yet to be observed in particle accelerators. If superpartners of elementary particles are detected at higher energy scales where supersymmetry is unbroken, then it would provide the engineer the opportunity to design quantum gates for computers by operating at higher energies.

Author

Table of Contents

Chapter 1

Supersymmetry

1.Introduction: In this chapter, we introduce the basic calculus of superfields, infinitesimal supersymmetry transformation and the construction of Lagrangians from superfields whose space-time integrals are super-symmetry invariant. Just as the Bosonic sector is described by four commuting space-time coordinates, the Fermionic sector is described by four anticommuting Majorana Fermionic coordinates. By Majorana Fermions, we mean anticommuting variables whose complex conjugates can also be expressed as linear combinations of themselves. Just as we have canonical commutation relations between the four Bosonic coordinates and the corresponding partial derivatives, so also we have canonical anticommutation relations between the four Fermionic coordinates and the corresponding partial derivatives. A superfield is a polynomial in the Fermionic coordinates whose coefficients are smooth functions of the Bosonic coordinates. This polynomial can be at most of degree four since a product of more than four mutually anticommuting variables must necessarily be zero. Just as infinitesimal Bosonic transformations (ie, infinitesimal generators of Bosonic transformations) can be expressed in terms of Bosonic vector fields which are first order linear partial differential operators in the Bosonic coordinates, so also an infinitesimal supersymmetry transformation is a super vector field expressible as a linear combination of first order Bosonic and Fermionic partial derivatives with the coefficients of the Fermionic part being constants and the coefficients of the Bosonic part being proportional to the Fermionic coordinates. When such a supervector field acts on a superfield, we get another superfield whose component functions are obtained as a linear partial differential operators acting on the original superfield components. Supervector fields satisfy canonical anticommutation relations, ie, their anticommutators are given by linear combinations of Bosonic space-time derivatives. Thus, by combining the Poincare Lie algebra with the supervector field, we are able to get a representation of the superalgebra of Bosonic and Fermionic operator in terms of vector fields and super vector fields. A super Lie algebra is simply a graded vector space with $\mathbb{Z}_2$ grading with a super Lie bracket defined by $[X, Y]_S = XY - (-1)^{p(X)p(Y)}YX$ such that the vector space is closed under this bracket. Here, $p(X)$ denotes the parity of X, ie, $p(X)$ is a zero or a one depending on whether X is even or odd. This bracket is extended by bilinearity to the entire vector space. A graded vector space is also called a super vector space. Using the representation provided by super vector fields of a super Lie algebra, we can develop a superdifferential calculus by introducing additional features such as left Chiral and right Chiral Fermions based on properties of the Dirac Gamma matrices. When the components of a supervector field are expressed using the Dirac Gamma matrices and the Pauli spin matrices, then these components acquire a natural interpretation in terms of well known physical fields like the scalar Klein-Gordon field, the Dirac field, the electromagnetic and non-Abelian gauge and matter fields, their superpartners and the auxiliary fields. Supersymmetric Lagrangians are functions constructed from the components of a superfield which change only by a total space-time divergence under infinitesimal supersymmetry transformations and hence the associated action functionals obtained by forming their space-time integrals are supersymmetry invariant. An infinitesimal supersymmetry transformation is a linear combination of two terms, the first multiplies a component field by a single Fermionic parameter (ie, increases the Fermionic parameter degree by one) and differentiates it w.r.t to space-time and the second simply lowers the Fermionic degree by one by differentiating it w.r.t the Fermionic parameters. Thus, infinitesimal supersymmetry transformations have odd parity and therefore transform Bosonic fields by Fermionic fields and vice versa. Thus, the invariance of an action under supersymmetry has the natural interpretation of being invariant under Boson-Fermion interchange. Associated with any symmetry of the action, is a Noether current that is conserved. In particular, associated with supersymmetry of a Lagrangian is a supersymmetry current and its conservation equation is a fundamental law of supersymmetry physics. In the superspace domain, it leads to the conservation law of a supercurrent that contains many additional conservation laws of physics like that of the energy-momentum tensor. Using the calculus of Chiral fields and the notion of a gauge transformation that leaves the Lagrangian invariant, one can naturally construct Lagrangians for matter and gauge fields that possess all the kinds of invariance that we desire, namely Lorentz invariance, gauge invariance and supersymmetry invariance. The equations of motion that follow from such a Lagrangian are natural generalizations of all the known laws of physics like Klein-Gordon scalar field equations, the Dirac relativistic wave equation for the electron and the electromagnetic and non-Abelian gauge field equations that describe the propagation of photons and nuclear forces and even gravity if one extends supersymmetry to local supersymmetry. The fascinating this about supersymmetry is that all these laws of physics are derived from a single superfield and are therefore related to each other.

The simplest model for a super Yang Mills field theory that possesses local supersymmetry invariance comprises of a Lagrangian that is the sum of a non-Abelian gauge field component in which the gauge potentials are Bosonic and a component that is just like a massless Dirac Lagrangian but with the corresponding field being the Fermionic superpartner of the Bosonic gauge potential. By writing down the supersymmetry transformation for such a Lagrangian, one obtains local supersymmetry only if the Dirac matrices appearing in this Lagrangian has certain dimensions like ten. Thus, we can construct a ten dimensional super-Yang-Mills theory in a simple way. This fact suggests very strongly that space-time in the Fermionic sector should have dimension ten. A major part of superstring theory is based on this principle.

Likewise the simplest model for supergravity comprise the sum of two terms. This first is the familiar Einstein-Hilbert action namely the scalar space-time curvature expressed in terms of the spinor connection of the gravitational field in tetrad notation. This tetrad is the graviton field having spin two. The second component is the gravitino field having spin 3/2 that is the superpartner of the graviton. This component has a Lagrangian that resembles that of the Dirac Lagrangian in curved space-time with the space-times partial derivatives being replaced by covariant derivatives relative to the spinor connection but with an antisymmetrized version of the product of three Dirac Gamma matrices appearing in place of a single Dirac Gamma matrix. Such a Lagrangian is seen to have local supersymetry provided that the supersymmetry transformation of the spinor connection is derived from its field equation which turns out to be algebraic.

When one wishes to describe both super-Yang Mills fields and supergravity, then additional field have to be introduced into the sum of these two components to maintqain supersymmetry and we have just indicated the reference source for this: M.Green, J.Schwarz and E.Witten, Superstring Theory, vol.2, Cambridge University Press. Supersymmetry can be broken by a superpotential having particular properties or by directly introducing external c-number control currents that couple to the gauge fields or by introucing c-number gauge potentials that couple to the matter fields. This is analogous to introducing a c-number four current density that interacts with the electromagnetic four potential or a c-number electromagnetic 4-vector potential that interacts with the Dirac four current density. When the Lagrangian is thus perturbed by c-number control signal, supersymmetry is broken. This is also just as one has in the electroweak theory a Yukawa coupling between the electron like leptons and the Higgs double field which when the Higgs field is replaced by its vacuum expectation value, the gauge symmetry is broken thereby giving the electron its mass. Once supersymmetry is broken we can control the c-number signals so that the corresponding perturbed Lagrangian describes via the Feynman path integral a quantum unitary gate, ie, a scattering matrix between an initial and a final state of the fields and by adjusting these control signals, we can match the scattering matrix to a given gate like the quantum Fourier transform gate.

[1] Application of supersymmetry to the design of quantum unitary gates. What is meant by a supersymmetric theory of elementary particles ?

[a] What are Majorana Fermionic anticommuting variables $\theta = (\theta^a : a = 1, 2, 3, 4)$?

[b] A superfield $S(x, \theta)$ is an arbitrary smooth function of the four Bosonic space-time coordinates $x = (x^\mu : \mu = 0, 1, 2, 3)$ and the four Fermionic coor-

dinates θ. It can be expressed as a fourth degree polyonmial in the Fermionic variables with coefficients being functions of the Bosonic variables.

[c] What are supersymmetry generators $\{Q_a, \bar{Q}_a : a = 0,1,2,3\}$? They satisfy the anticommutation relations

$$\{Q_a, \bar{Q}_b\} = \gamma^\mu_{ab} P_\mu$$

where P_μ is the Bosonic four momentum. Here

$$\bar{Q} = Q^T \gamma^5 \epsilon, \gamma^5 = [I_2, -I_2], \epsilon = diag[i\sigma_y, -i\sigma_y]$$

A realization of supersymmetry generators is provided by super-vector fields

$$L_a = (\gamma^\mu \theta)_a \partial/\partial x^\mu + (\gamma^5 \epsilon)_{ab}.\partial/\partial\theta^b$$

and

$$\bar{L}_a = (\gamma^5 \epsilon)_{ab} L_b$$

Show using the Fermionic anticommutation analog of the Bosonic commutation relations

$$[\partial/\partial x^\mu, x^\nu] = \delta^\nu_\mu$$

namely,

$$\{\partial/\partial\theta^a, \theta^b\} = \delta^b_a$$

that

$$\{L_a, \bar{L}_b\} = \gamma^\mu_{ab} \partial/\partial x^\mu$$

Note that

$$P_\mu = i\partial/\partial x^\mu$$

and hence the operators $\{L_a, \bar{L}_a\}$ provide us with a representation of the supersymmetric Lie algebra generated by $\{J_{\mu\nu}, P_\mu, Q_a, \bar{Q}_a\}$ where $J_{\mu\nu}$ are the four angular momentum operators

$$J_{\mu\nu} = x_\mu P_\nu - x_\nu P_\mu$$

Note that $\{J_{\mu\nu}, P_\mu\}$ generate the Poincare Lie algebra consisting of Lorentz transformations and space-time translations. So here by incorporating Fermionic operators into this Lie algebra, we obtain the super-Poincare Lie algebra.

[d] Supersymmetric current and its conservation: Let $S(x, \theta)$ be a superfield and let $\chi_m(x), m = 1, 2, ...$ denote its component fields. Let $\mathcal{L}$ be a supersymmetric Lagrangian constructed from these component fields. Under an infinitesimal supersymmetry transformation, $\mathcal{L}$ changes by a total four space-time divergence, ie,

$$\delta\mathcal{L} = \partial_\mu J^\mu$$

We can also associate a Noether current N^μ associated with with the change in the Lagrangian under this infinitesimal supersymmetric transformation of the component fields:

$$N^\mu = \frac{\partial\mathcal{L}}{\partial\chi_{m,\mu}} \delta\chi_m$$

The Noether current is conserved when the field equations are satisfied provided that the Lagrangian is invariant under the supersymetry transformation follows immediately by making use of the Euler-Lagrange equations. We have in fact, from the Euler-Lagrange equations,

$$\partial_\mu \frac{\partial \mathcal{L}}{\partial \chi_{m,\mu}} = \frac{\partial \mathcal{L}}{\partial \chi_m}$$

and therefore,

$$\partial_\mu N^\mu = (\partial \mathcal{L}/\partial \chi_m)\delta\chi_m + \frac{\partial \mathcal{L}}{\partial \chi_{m,\mu}}\delta\chi_{m,\mu} = \delta\mathcal{L}$$

provided that the equations of motion are satisfied. This is zero only when the Lagrangian is invariant under a supersymmetry transformation. However, in general, under a supersymmetry transformation as we noted above, the Lagrangian changes by a four divergence and is not invariant. In other words, only the action integral is supersymmetry invariant. Thus, in general, we can only write

$$\partial_\mu N^\mu = \delta L$$

provided that the equations of motion are satisfied. It follows that the difference of the two equations gives a conservation law:

$$\partial_\mu (S^\mu - N^\mu) = 0$$

in the general case when the equations of motion are satisfied.

Remark: $\partial_\mu J^\mu = \delta L$ is always true while $\partial_\mu N^\mu = \delta L$ is true only when the equations of motion are satisfied. Therefore, in particular, both the equations are valid when the equations of motion are satisfied.

[e] Left and right superderivatives:

$$\theta_L = (1+\gamma^5)\theta/2, \theta_R = (1-\gamma^5)\theta/2$$

Then,

$$\begin{aligned} D &= \gamma^\mu \theta.\partial_\mu - \gamma^5 \epsilon \partial_\theta \\ D_L = (1+\gamma^5)D/2 &= \gamma^\mu \theta_R \partial_\mu - \epsilon \partial_{\theta_L} \\ &= \gamma^\mu \theta_R \partial_\mu - \gamma^5 \epsilon \partial_{\theta_L} \\ D_R = (1-\gamma^5)D/2 &= \gamma^\mu \theta_L \partial_\mu + \epsilon \partial_{\theta_R} \\ &= \gamma^\mu \theta_L \partial_\mu - \gamma^5 \epsilon \partial_{\theta_R} \end{aligned}$$

Note that

$$\gamma^5 \gamma^\mu = -\gamma^\mu \gamma^5$$

Further, it is clear that

$$(1-\gamma^5).D_L = 0, (1+\gamma^5)D_R = 0$$

and hence only two of the $D_L's$ are linearly independent and likewise only two of the $D_R's$ are independent. Further any two of the $D_L's$ anticommute and also any two of the $D_R's$ anticommute:

$$\{D_L, D_L^T\} = \{D_R, D_R^T\} = 0$$

because any two $\theta_R's$ anticommute, any two $\partial_{\theta_L}'s$ anti commute and θ_R anticommutes with ∂_{θ_L} and likewise, θ_L anticommutes with ∂_{θ_R}. Since any two of the $D_R's$ anticommute and since there are only two linearly independent $D_R's$, it follows easily that the product of any three or more $D_R's$ is zero and likewise, the product of any three or more $D_L's$ is also zero.

Further,

$$\{\theta_L, \partial_{\theta_L}^T\} = \{\partial_{\theta_L}, \theta_L^T\} = (‘1 + \gamma^5)/2,$$

and likewise

$$\{\theta_R, \partial_{\theta_R}^T\} = \{\partial_{\theta_R}, \theta_R^T\} = (1 - \gamma^5)/2,$$

It follows that

$$\{D_L, D_R^T\} = \{\gamma^\mu \theta_R \partial_\mu - \epsilon \partial_{\theta_L}, (\gamma^\mu \theta_L \partial_\mu + \epsilon \partial_{\theta_R})^T\}$$

$$= [-\gamma^\mu (1 - \gamma^5)\epsilon/2 - \epsilon(1 + \gamma^5)\gamma^{\mu T}/2]\partial_\mu$$

$$= [-\gamma^\mu (1 - \gamma^5)\epsilon/2 - \gamma^\mu \epsilon(1 - \gamma^5)/2]\partial_\mu$$

$$= -\gamma^\mu (1 - \gamma^5)\epsilon \partial_\mu$$

This equation is the same as

$$\{D_{La}, D_{Rb}\} = -[\gamma^\mu (1 - \gamma^5)\epsilon]_{ab} \partial_\mu$$

which is equivalent to

$$\{D_{Ra}, D_{Lb}\} = -[\gamma^\mu (1 - \gamma^5)\epsilon]_{ba} \partial_\mu$$

or equivalently,

$$\{D_R, D_L^T\} = -[\gamma^\mu (1 - \gamma^5)\epsilon]^T \partial_\mu$$

$$= \epsilon(1 - \gamma^5)\gamma^{\mu T} \partial_\mu$$

$$= \epsilon \gamma^{\mu T}(1 + \gamma^5)\partial_\mu = \gamma^\mu \epsilon(1 + \gamma^5)\partial_\mu$$

A left Chiral superfield is by definition any function of θ_L and

$$x_+^\mu = x^\mu + \theta_R^T \epsilon \gamma^\mu \theta_L$$

We prove that a superfield Φ is left Chiral iff $D_R \Phi = 0$. The necessity part will follow if we can show that $D_R \theta_L = 0$ and $D_R x_+^\mu = 0$. But

$$D_R \theta_L = [\gamma^\mu \theta_L \partial_\mu + \epsilon \partial_{\theta_R}]\theta_L = 0$$

since

$$\partial_\mu \theta_L = 0 and \partial_{\theta_R} \theta_L = 0$$

Also,

$$D_R x_+^\nu = [\gamma^\mu \theta_L \partial_\mu + \epsilon \partial_{\theta_R}](x^\nu + \theta_R^T \epsilon \gamma^\nu \theta_L)$$

$$= \gamma^\nu \theta_L + \epsilon \partial_{\theta_R}(\theta_R^T \epsilon \gamma^\nu \theta_L$$

$$= \gamma^\nu \theta_L + \epsilon^2 \gamma^\nu \theta_L = 0$$

since

$$\epsilon = diag[e, e], e = i\sigma^2$$

implies

$$\epsilon^2 = -I_4$$

To prove the converse, define

$$x_-^\mu = x^\mu - \theta_R^T \epsilon \gamma^\mu \theta_L$$

Then, is clear that any superfield can be expressed as a function of $\theta_L, \theta_R, x_+^\mu and x_-^\mu$. So it suffices to prove that

$$D_R \theta_R, D_R x_-^\mu$$

are no-zero. This follows at once.

[f] Superfield equations: We know that if $S(x, \theta)$ is a superfield, then under a supersymmetry transformation, $[S]_D$ changes by a four space-time divergence and hence the integral $\int [S]_D d^4x$ is supersymmetry invariant, ie,

$$\int [\bar{\alpha} L S] d^4x = 0$$

where $\bar{\alpha} = \alpha^T \gamma^5 \epsilon$ with α a Majorana Fermionic parameter. Now it is well known that the class of left Chiral superfields is invariant under a supersymmetry transformation and so is the class of right Chiral superfields. To see this, we need only note that

$$L(\theta_L)^T = (\gamma^\mu \theta \partial_\mu + \gamma^5 \epsilon \partial_\theta) \theta_L^T$$

$$= \gamma^5 \epsilon (1 + \gamma^5)/2 = \epsilon (1 + \gamma^5)/2$$

which is a constant matrix and further,

$$L x_+^\nu = (\gamma^\mu \theta \partial_\mu + \gamma^5 \epsilon \partial_\theta)(x^\nu + \theta_R^T \epsilon \gamma^\nu \theta_L)$$

$$= \gamma^\nu \theta + \gamma^5 \epsilon \partial_\theta \theta_R^T \epsilon \gamma^\nu \theta_L$$

$$= \gamma^\nu \theta + \gamma^5 \epsilon ((1 - \gamma^5) \epsilon \gamma^\nu \theta_L + (1 + \gamma^5) \epsilon \gamma^\nu \theta_R)/2$$

$$= \gamma^\nu \theta + \gamma^5 \epsilon . \epsilon \gamma^\nu (\theta_L + \theta_R)$$

$$= 0$$

thus proving the claim. Note that we have used the facts that

$$\theta = \theta_L + \theta_R,$$

$$(\epsilon\gamma^\nu)^T = -\epsilon\gamma^\nu$$

and hence, since θ_L and θ_R anticommute,

$$\theta_R^T \epsilon\gamma^\nu \theta_L = \theta_L^T \epsilon\gamma^\nu \theta_R$$

[g] Construction of supersymmetric Lagrangians from Chiral superfields.

Let Φ be left invariant Chiral field. We claim that if $[\Phi]_F$ denotes the coefficient of $\theta_L^T \epsilon\theta_L$ in Φ, then $[\Phi]_F$ changes by a four spatio-temporal divergence under a supersymetry transformation. To see this, using the left Chiral property of Φ,we expand Φ as

$$\Phi(\theta_L, x_+^\mu) = \phi_1(x_+) + \theta_L^T \epsilon\phi_2(x_+)$$

$$+\theta_L^T \epsilon\theta_L.\phi_3(x_+)$$

Note that product of three or more of the $\theta_L' s$ is zero and hence the last term is

$$\theta_L^T \epsilon\theta_L.\phi_3(x_+) = \theta_L^T \epsilon.\phi_3(x)$$

Now applying the infinitesimal supersymmetry transformation $\bar{\alpha}L$ to Φ gives

$$\alpha^T \gamma^5 \epsilon.L\Phi$$

and we find that

$$L\Phi = (\gamma^\mu \theta\partial_\mu + \gamma^5 \epsilon.\partial_\theta)\Phi$$

and we easily see that the term in this expression that is quadratic in the $\theta' s$ is given by

$$\gamma^\mu \theta.(\theta_L^T \epsilon\phi_{2,\mu}(x)) +$$

$$+\gamma^5 \epsilon.\partial_\theta(\theta_L^T \epsilon.\phi_{2,\mu}(x)\theta_R^T \epsilon\gamma^\mu \theta_L)$$

and it is easily seen that this is a perfect four space-time divergence and in particular, the coefficient of $\theta_L^T \epsilon\theta_L$ in a perfect space-time four divergence. This proves the claim. Now, if Φ is left Chiral, then for any function f of one variable, $f(\Phi)$ (to be interpreted as a power series in Φ is again left Chiral, since it is also a function of θ_L and x_+^μ only and hence $[f(\Phi)]_F$ is a candidate for a supersymmetric Lagrangian. If $K(x,y)$ is a function of two variables and Φ is left Chiral, then $[K(\Phi^*, \Phi)]_D$ is also a candidate for a supersymmetric Lagrangian. Thus we take for our matter field supersymmetric Lagrangian

$$L_M = [K(\Phi^*, \Phi)]_D + [f(\Phi)]_F$$

In quantum field theory, the matter fields are composed of scalar Klein-Gordon fields, the Dirac electron-positron field and the Yang-Mills matter field the latter is an extended version of the Dirac electron-positron field and includes particles like nucleons.

Remark: The construction of conserved currents and the associated symmetry generated by the corresponding charge is familiar even in classical mechanics. For example, suppose $L(q, q')$ is a Lagrangian which changes by a total time derivative under the infinitesimal symmetry $q \rightarrow q + \delta q(q, q')$. Then we can write

$$(\partial L/\partial q)\delta q + (\partial L/\partial q')\delta q' = dF(q, q')/dt$$

$$= (\partial F/\partial q)q' + (\partial F/\partial q')q''$$

Here, we are assuming that F depends only on q, q'. Now,

$$\delta q' = (\partial \delta q/\partial q)q' + (\partial \delta q/\partial q')q''$$

and hence equating the coefficients of q'' on both sides gives

$$(\partial L/\partial q')(\partial \delta q/\partial q') = \partial F/\partial q' - - - (1a)$$

or equivalently,

$$p\partial \delta q/\partial q' = \partial F/\partial q' - - - (1b)$$

and thus, we also have

$$(\partial L/\partial q)\delta q + (\partial L/\partial q')(\partial \delta q/\partial q)q'$$

$$= (\partial F/\partial q)q' - - - (2a)$$

or equivalently,

$$(\partial L/\partial q)\delta q + p(\partial \delta q/\partial q)q'$$

$$= (\partial F/\partial q)q' - - - (2b)$$

When the Euler-Lagrange equations of motion are satisfied, it then follows that

$$d/dt((\partial L/\partial q')\delta q - F) =$$

$$d/dt(\partial L/\partial q')\delta q + (\partial L/\partial q')((\partial \delta q/\partial q)q' + (\partial \delta q/\partial q')q'') - dF/dt$$

$$= (\partial L/\partial q)\delta q + (\partial L/\partial q')((\partial \delta q/\partial q)q' + (\partial \delta q/\partial q')q'') - dF/dt = 0$$

In fact, this conservation law does not depend on using the fact that F is a function of only q, q'. It can be any function of time. However, suppose we introduce the conserved charge

$$Q = (\partial L/\partial q')\delta q - F = p\delta q - F$$

We just showed that when the equations of motion are satisfied,

$$Q' = 0$$

We now assume that $F = F(q, q')$. Then $Q = Q(q, q')$ and therefore Q is an observable in the context of classical mechanics. Further, by virtue of (1),

$$\{Q, q\} = \{p, q\}\delta q + p\{\delta q, q\} - \{F, q\}$$

$$= \delta q + p\partial\delta q/\partial q'\{q', q\} - (\partial F/\partial q')\{q', p\}$$

$$= \delta q + (p\partial\delta q/\partial q' - \partial F/\partial q')\{q', p\}$$

$$= \delta q$$

This equation is true under all conditions, ie, even when the equations of motion are not satisfied. Further, under the conditions that the equations of motion are satisfied,

$$\delta q' = \{Q, q\}' = \{Q', q\} + \{Q, q'\} = \{Q, q'\}$$

Thus, indeed Q generates the symmetry group of transformations that leave the action integral invariant. In the context of fields and supersymmetry, the supesymmetric transformation changes $\mathcal{L}$ by a four space-time divergence:

$$\delta L = \partial_\mu J\mu$$

Here, the role played by F in the above discussion on particle mechanics is played by J^μ and the role played by dF/dt is played by $\partial_\mu J^\mu$. The role played by

$$(\partial L/\partial q)\delta q + (\partial L/\partial q')\delta q'$$

which when the equations of motion are satisfied, equals

$$d/dt((\partial L/\partial q')\delta q)$$

is played by

$$\partial_\mu N^\mu$$

In other words, the role played by

$$p\delta q = (\partial L/\partial q')\delta q$$

is played by N^μ. Thus the role played by the conserved charge $Q = p\delta q - F$ is played by the supersymmetry current $N^\mu - J^\mu$. The role played by the Noether conservation of charge equation

$$Q' = 0$$

is played by the conservation equation of supersymmetry current

$$\partial_\mu(N^\mu - J^\mu) = 0$$

If Φ is a left Chiral superfield, then for any function f of one variable defined as a power series, $f(\Phi)$ is also Left Chiral and hence $[f(\Phi)]_F$ is a supersymmetric

Lagrangian and so is $[\Phi^*\Phi]_D$. A candidate Lagrangian for the matter field is then

$$L_M = [\Phi^*\Phi]_D + c_1[f(\Phi)]_F$$

Now, Φ can be expressed as $D_R^2 S$ where S some superfield and $D_R^2 = \bar{D}_R\epsilon D_R$ with

$$\bar{D}_R = \bar{D}(1-\gamma^5)/2 = D^T\gamma^5\epsilon(1-\gamma^5)/2$$
$$= D_R^T\gamma^5\epsilon$$

Recall that since the product of any three $D_R's$ is zero, it follows that $D_R^2 S$ is annihilated by D_R and hence is Left Chiral. Now

$$[f(\Phi)]_F = [f(D_R^2 S)]_F$$

combined with the fact that D_R^2 equals $\partial_{\theta_R}^T\epsilon\partial_{\theta_R}$ plus terms involving the Bosonic derivatives ∂_μ with constant Fermionic parameters implies (using the power series expansion of f that $f(D_R^2 S)$ equals $D_R^2\tilde{S}$ plus perfect Bosonic divergences for some superfield $\tilde{S}$. Note for example that

$$(D_R^2 S)^2 = D_R^2(S.D_R^2 S) + X$$

where X where X is a perfect Bosonic divergence with constant Fermionic coefficients. Now, $[D_R^2\tilde{S})]_F$ which is the coefficient of $\theta_L^T\epsilon\theta_L$ in $D_R^2\tilde{S}$, must coincide with a nonzero real constant c_2 times $[\tilde{S}]_D$ plus a total space-time four divergence. Note that $\theta_L^T\epsilon\theta_L.\theta_R^T\epsilon\theta_R$ equals a non-zero real constant times $(\theta^T\epsilon\theta)^2$. Recall that $[S]_D$ equals the coefficient of $(\theta^T\epsilon\theta)^2$ in S plus a perfect space-time divergence. Hence, in terms of the Berezin integral, we can write

$$\int [f(\Phi)]_F d^4x = \int [D_R^2\tilde{S}]_F d^4x = c_2\int [\tilde{S}]_D d^4x = c_2\int \tilde{S}d^4x d^4\theta$$

Now, we can write the total matter action as

$$\int (D_R^2 S)^*(D_R^2 S)d^4x d^4\theta + c_1\int [f(D_R^2 S)]_F d^4x$$

We recall that from the definition of Majorana Fermionic parameters,

$$\theta^* = \gamma^5\epsilon\gamma^0\theta$$

where $\gamma^5\epsilon\gamma^0$ has the block structure

$$\begin{pmatrix} 0 & e \\ -e & 0 \end{pmatrix}$$

and, we can interpret this equation in terms of 2×1 components as

$$\theta_L^* = e\theta_R, \theta_R^* = -e\theta_L$$

where

$$e^2 = -I_2$$

This is in agreement with the condition

$$(\theta_L^*)^* = \theta_L, (\theta_R^*)^* = \theta_R$$

since

$$e^* = e, e^2 = -I_2$$

provided that we interpret θ_L and θ_R as 2×1 Fermionic vectors. Note that

$$\epsilon = diag[e, e], \gamma^5 \epsilon = diag[e, -e]$$

we can write

$$\bar{\theta} = (\theta^*)^T \gamma^0 = \theta^T \gamma^5 \epsilon$$

and hence,

$$\bar{\theta}_L = \theta_L^T \gamma^5 \epsilon, \bar{\theta}_R = \theta_R^T \gamma^5 \epsilon$$

when interpreted as 4×1 Fermionic vectors. Also,

$$D^* = \gamma^5 \epsilon \gamma^0 D$$

so that

$$\bar{D} = (D^*)^T \gamma^0 = -D^T \gamma^0 \gamma^5 \epsilon \gamma^0 = D^T \gamma^5 \epsilon$$

and hence,

$$D_R^* = ((1 - \gamma^5)/2) D^* = \gamma^5 \epsilon \gamma^0 D_L,$$
$$D_L^* = ((1 + \gamma^5)/2) D^* = \gamma^5 \epsilon \gamma^0 D_R$$

when interpreted as 4×1 vector valued super-vector fields. When interpreted as 2×1 vector valued super vector fields, these equations should be read as

$$D_R^* = -eD_L, D_L^* = eD_R$$

when interpreted as 2×1 vector fields, it should be understood that D_L is represented by $\begin{pmatrix} D_L \\ 0 \end{pmatrix}$ and D_R by $\begin{pmatrix} 0 \\ D_R \end{pmatrix}$. Then,

$$\int [\Phi^* \Phi]_D d^4x = \int \Phi^* \Phi d^4x d^4\theta$$

$$= \int (D_R^2 S)^* (D_R^2 S) d^4x d^4\theta = \int \Phi^* D_R^2 S d^4x d^4\theta$$

$$= -\int (D_L^2 \Phi^*) S d^4x d^4\theta$$

This is because

$$\Phi^* D_R^2 S = \Phi^* D_R^T \epsilon D_R S =$$
$$(-1)^{p(\Phi)} D_R^T (\Phi^* \epsilon D_R S) - (1)^{p(\Phi)} D_R^T \Phi^* . \epsilon D_R S$$
$$= (-1)^{p(\Phi)} D_R^T (\Phi^* \epsilon D_R S) + (1)^{p(\Phi)} (\epsilon D_R)^T \Phi^* . D_R S$$

$$= (-1)^{p(\Phi)} D_R^T(\Phi^* \epsilon D_R S) + D_R^T((\epsilon D_R)\Phi^*.S)$$
$$-(D_R^T \epsilon D_R \Phi^*).S$$

and

$$-(D_R^T \epsilon D_R \Phi^*)S == -(D_R^{*T} \epsilon D_R^* \Phi)^* S$$
$$= -(D_L^T (\gamma^5 \epsilon \gamma^0)^T \epsilon \gamma^5 \epsilon \gamma^0 D_L \Phi)^* S$$
$$= (D_L^T \gamma^0 \epsilon \gamma^5 \epsilon \gamma^5 \epsilon \gamma^0 D_L \Phi)^* S$$
$$= -(D_L^T \gamma^0 \epsilon \gamma^0 D_L \Phi)^* S =$$
$$= -(D_L^T \epsilon D_L \Phi)^* S = -(D_L^2 \Phi)^* S$$

Thus,

$$\int \Phi^* D_R^2 S d^4x d^4\theta = -\int (D_L^2 \Phi)^* S d^4x d^4\theta$$

and in exactly the same way, we can show that

$$\int [\Phi^* \delta\Phi]_D d^4x = \int [\Phi^* D_R^2 \delta S]_D d^4x$$

$$= \int \Phi^* D_R^2 \delta S d^4x d^4\theta = -\int (D_L^2 \Phi)^* \delta S d^4x d^4\theta$$

Likewise consider

$$\delta \int [f(D_R^2 S)]_F d^4x = \int [\delta f(D_R^2 S)]_F d^4x$$

Now consider for example

$$\delta(D_R^2 S) = D_R^2 \delta S,$$
$$\delta((D_R^2 S)^2) = D_R^2 \delta S.D_R^2 S + (-1)^{p(S)} D_R^2 S.D_R^2 \delta S$$
$$= D_R^2 \delta S.D_R^2 S + (-1)^{p(S)+p(S)p(\delta S)} + D_R^2 \delta S.D_R^2 S$$
$$= 2 D_R^2 \delta S.D_R^2 S$$

since

$$p(\delta S) = p(S)$$

Likewise, in general, we have

$$\delta(D_R^2 S)^n = D_R^2 \delta S.n(D_R^2 S)^{n-1}$$

and hence,

$$\delta f(D_R^2 S) = D_R^2 \delta S.f'(D_R^2 S) = D_R^2 \delta S.f'(\Phi)$$

Now,

$$D_R f'(\Phi) = D_R f'(D_R^2 S) = 0$$

since D_R operating on a constant is zero and the product of three $D_R's$ is zero. Thus,

$$D_R(\delta S.f'(\Phi)) = D_R \delta S.f'(\Phi),$$

$$D_R^2(\delta S.f'(\Phi)) = D_R^2\delta S.f'(\Phi)$$

Thus,

$$\delta\int[f(\Phi)]_F d^4x = \int[D_R^2\delta S.f'(\Phi)]_F d^4x =$$

$$\int[D_R^2(\delta S.f'(\Phi))]_F d^4x = \int[\delta S.f'(\Phi)]_D d^4x$$

$$= \int f'(\Phi)\delta S d^4x d^4\theta$$

since δS, S and $D_R^2 S$ have the same parity. Thus, our superfield equations

$$\delta[\int[\Phi^*\Phi]_D d^4x + c_1\int[f(\Phi)]_F d^4x] = 0$$

for Φ a left Chiral field result in the superfield equations

$$-(D_L^2\Phi)^* + c_1 f'(\Phi) = 0$$

or equivalently

$$D_L^2 D_R^2 S = \bar{c}_1 f'(D_R^2 S)^*$$

Note that both sides of this equation are right Chiral superfields. More generally, if we replace the term $[\Phi^*\Phi]_D$ in the Lagrangian by $[K(\Phi^*, \Phi)]_D$ with $\Phi = D_R^2 S$, then the variation of the corresponding action integral w.r.t S is given by

$$\int[\delta K((D_R^2 S)^*, D_R^2 S)/\delta\Phi]D_R^2\delta S d^4x d^4\theta$$

$$= -\int[D_R^2\delta K(D_L^2 S^*, D_R^2 S)/\delta\Phi]\delta S d^4x d^4\theta$$

Note that in deriving this equation, we have used the fact that D_R acting on any superfield consists of a sum of terms having lesser that four Fermionic parameter products and terms that are total Bosonic space-time divergences. These terms cancel out when integrated over $d^4xd^4\theta$. So our superfield equations in this case generalize to

$$-D_R^2\delta K(D_L^2 S^*, D_R^2 S)/\delta\Phi + c_1 f'(D_R^2 S) = 0$$

Both sides here are left Chiral superfields.

[h] Supergravity:

Let ω_μ^{mn} denote the spinor connection of the gravitational field. Then if Γ_m are the Dirac matrices in four dimensions and e_μ^m is the tetrad basis of space time being used, the covariant derivative of a spinor field is defined by

$$D_\mu\psi = (\partial_\mu + (1/4)\omega_\mu^{mn}\Gamma_{mn})\psi$$

where

$$\Gamma_{mn} = [\Gamma_m, \Gamma_n]$$

The curvature tensor in spinor notation is

$$R_{\mu\nu} = [\partial_\mu + (1/4)\omega_{\mu\nu}^{mn}\Gamma_{mn}, \partial_\nu + (1/4)\omega_\nu^{rs}\Gamma_{rs}]$$

$$= (1/4)(\omega_{\nu,\mu}^{mn} - \omega_{\nu,\mu}^{mn})\Gamma_{mn}$$

$$+(1/16)\omega_\mu^{mn}\omega_\nu^{rs}[\Gamma_{mn}, \Gamma_{rs}]$$

Now using the anticommutator

$$\{\Gamma_m, \Gamma_n\} = 2\eta_{mn}$$

we can easily show that

$$[\Gamma_{mn}, \Gamma_{rs}] = 4(\eta_{ms}\Gamma_{nr} + \eta_{nr}\Gamma_{ms} - \eta_{mr}\Gamma_{ns} - \eta_{ns}\Gamma_{mr})$$

Thus

$$R_{\mu\nu} =$$

$$= (1/4)(\omega_{\nu,\mu}^{mn} - \omega_{\nu,\mu}^{mn})\Gamma_{mn}$$

$$+(1/4)\omega_\mu^{mn}\omega_\nu^{rs}(\eta_{ms}\Gamma_{nr} + \eta_{nr}\Gamma_{ms} - \eta_{mr}\Gamma_{ns} - \eta_{ns}\Gamma_{mr})$$

This can be expressed as

$$R_{\mu\nu} = (1/4)R_{\mu\nu}^{mn}\Gamma_{mn}$$

where

$$R_{\mu\nu}^{mn} = \omega_{\nu,\mu}^{mn} - \omega_{\nu,\mu}^{mn} - \omega_\mu^{rn}\omega_\nu^{ms}\eta_{rs} + \omega_\mu^{ms}\omega_\nu^{rn}\eta_{sr} + \omega_\mu^{sn}\omega_\nu^{rm}\eta_{sr} - \omega_\mu^{mr}\omega_\nu^{ns}\eta_{rs}$$

$$=$$

$$\omega_{\nu,\mu}^{mn} - \omega_{\nu,\mu}^{mn} + \eta_{rs}(-\omega_\mu^{rn}\omega_\nu^{ms} + \omega_\mu^{ms}\omega_\nu^{rn} + \omega_\mu^{sn}\omega_\nu^{rm} - \omega_\mu^{mr}\omega_\nu^{ns})$$

$$= \omega_{\nu,\mu}^{mn} - \omega_{\nu,\mu}^{mn} + 2\eta_{rs}(\omega_\mu^{mr}\omega_\nu^{sn} - \omega_\nu^{mr}\omega_\mu^{ns})$$

It is easily shown that when the spinor connection ω_μ^{mn} for the gravitational field is appropriately chosen so that the Dirac equation in curved space-time remains invariant under both diffeormophisms and local Lorentz transformations, then the Riemann curvature tensor as defined usually in terms of the Christoffel connection symbols, coincides with $R_{\mu\nu\rho\sigma} = R_{\mu\nu}^{mn}e_{m\rho}e_{n\sigma}$. In particular, $R = R_{\mu\nu}^{mn}e_m^\mu e_n^\nu$ is the scalar curvature of space-time. The spinor connection ω_μ^{mn} is chosen so that the covariant derivative of the tetrad e_μ^n having one spinor index and one vector index is zero:

$$0 = D_\nu e_\mu^n = e_{\mu,\nu}^n - \Gamma_{\mu\nu}^\alpha e_\alpha^n + \omega_\nu^{nm}e_{m\mu}$$

This is an algebraic equation for ω_μ^{mn} and is easily solved. However when there are spinor fields like the gravitino in addition to the gravitational field specified by the tetrad e_n^μ (ie, the graviton), then the definition of the spinor connection has to be modified and it is expressed in terms of both the graviton and the gravitino fields. This equation is obtained by first considering the supergravity Lagrangian in four space-time dimensions

$$c_1 eR + i\bar{\chi}_\mu \Gamma^{\mu\nu\rho} D_\nu \chi_\rho$$

where χ_μ is a Majorana spinor having an additional vector index μ. The graviton tetrad field e_n^μ is Bosonic while the gravitino χ_μ is Fermionic. These are all considered in the quantum theory to be operator valued fields. Note the following self consistent definitions:

$$\Gamma^\mu = e_n^\mu \Gamma^n, \Gamma_\mu = g_{\mu\nu}\Gamma^\nu = \Gamma^n e_{n\mu}$$

$$\Gamma_n = \eta_{nm}\Gamma^m, e_\mu^n e_{n\nu} = g_{\mu\nu}, e_n^\mu e_{m\mu} = \eta_{nm},$$

$$e_{n\mu} = \eta_{nm} e_\mu^m = g_{\mu\nu} e_n^\nu$$

$$\Gamma^{\mu\nu\rho} = \Gamma^\mu \Gamma^{\nu\rho} + \Gamma^\nu \Gamma^{\rho\mu} + \Gamma^\rho \Gamma^{\mu\nu}$$

where

$$\Gamma^{\mu\nu} = [\Gamma^\mu, \Gamma^\nu]$$

Thus $\Gamma^{\mu\nu\rho}$ is obtained by antisymmetrizing the product $\Gamma^\mu\Gamma^\nu\Gamma^\rho$ over all the three indices. we can also clearly write

$$\Gamma^{\mu\nu} = e_m^\mu e_n^\nu \Gamma^{mn}, \Gamma^{\mu\nu\rho} = e_m^\mu e_n^\nu e_k^\rho \Gamma^{mnk}$$

In general, we can define

$$\Gamma^{\mu_1}..\Gamma^{\mu_k} = \sum_{\sigma \in S_n} sgn(\sigma)\Gamma^{\mu_{\sigma 1}}...\Gamma^{\mu_{\sigma k}}$$

This is obtained by totally antisymmetrizing the product $\Gamma^{\mu_1}...\Gamma^{\mu_k}$ over all its k indices. The basic property of a Majorana Fermionic operator field $\psi(x)$ is that apart from all its components anticommuting with each other, it has four components and satisfies

$$(\psi(x)^*)^T = \psi^T \Gamma^5 \epsilon \Gamma^0$$

where if

$$\psi(x) = \begin{pmatrix} \psi_1(x) \\ \psi_2(x) \\ \psi_3(x) \\ \psi_4(x) \end{pmatrix}$$

then

$$\psi(x)^* = \begin{pmatrix} \psi_1(x)^* \\ \psi_2(x)^* \\ \psi_3(x)^* \\ \psi_4(x)^* \end{pmatrix}$$

$\psi_k(x)^*$ denoting the operator adjoint of $\psi_k(x)$ in the Fock space on which it acts. Also we define

$$\psi(x)^T = [\psi_1(x), \psi_2(x), \psi_3(x), \psi_4(x)]$$

so that we have

$$(\psi(x)^*)^T = [\psi_1(x)^*, \psi_2(x)^*, \psi_3(x)^*, \psi_4(x)^*]$$

Now observe that

$$\epsilon = diag[i\sigma^2, i\sigma^2], \sigma^2 = \begin{pmatrix} 0 & -i \\ i & 0 \end{pmatrix}$$

Note that ϵ is a real skewsymmetric matrix. we write

$$e = i\sigma^2 = \begin{pmatrix} 0 & 1 \\ -1 & 0 \end{pmatrix}$$

so that

$$\epsilon = diag[e, e], e^2 = -I$$

$$\Gamma^5 \epsilon \Gamma^0 = \begin{pmatrix} e & 0 \\ 0 & -e \end{pmatrix} \begin{pmatrix} 0 & I \\ I & 0 \end{pmatrix}$$

$$= \begin{pmatrix} 0 & e \\ -e & 0 \end{pmatrix}$$

Thus, the condition for ψ to be a Majorana Fermion can be stated as

$$\psi_{1:2}^* = e\psi_{3:4}, \psi_{3:4}^* = -e\psi_{1:2}$$

or equivalently,

$$(\psi_{1:2}^*)^T = -(\psi_{3:4}^*)^T e, (\psi_{3:4}^*)^T = (\psi_{1:2}^*)^T e$$

Also,

$$\Gamma^0\Gamma^n, \Gamma^0\Gamma^\mu, \Gamma^n\Gamma^0, \Gamma^\mu\Gamma^0$$

are Hermitian matrices. We observe that if ψ is a Majorana Fermion,

$$\bar{\psi} = (\psi^*)^T\Gamma^0 = \psi^T\Gamma^5\epsilon\Gamma^0$$

So we can also write down the Lagrangian of the gravitino as

$$i\bar{\chi}_\mu \Gamma^{\mu\nu\rho} D_\nu \chi_\rho$$

$$= i\chi_\mu^{*T}\Gamma^0\Gamma^{\mu\nu\rho} D_\nu \chi_\rho$$

$$= i\chi_\mu^T \Gamma^5 \epsilon \Gamma^0 \Gamma^{\mu\nu\rho} D_\nu \chi_\rho$$

We can verify that apart from a perfect divergence, this quantity is a Hermitian operator field. First observe that

$$(\Gamma^0\Gamma^\mu\Gamma^\nu\Gamma^\rho)^* =$$

$$(\Gamma^0\Gamma^\mu\Gamma^\nu\Gamma^0\Gamma^0\Gamma^\rho)^* =$$

$$\Gamma^0\Gamma^\rho\Gamma^\nu\Gamma^0\Gamma^0\Gamma^\mu$$

$$= \Gamma^0\Gamma^\rho\Gamma^\nu\Gamma^\mu$$

so that on antisymmetrizing over the three indices, we get

$$(\Gamma^0\Gamma^{\mu\nu\rho})^* = -\Gamma^0\Gamma^{\mu\nu\rho}$$

Thus,

$$(i\chi_\mu^{*T}\Gamma^0\Gamma^{\mu\nu\rho}\partial_\nu\chi_\rho)^*$$

$$= i\chi_{\rho,\nu}^{*T}\Gamma^0\Gamma^{\mu\nu\rho}\chi_\mu$$

$$= i\chi_{\mu,\nu}^{*T}\Gamma^0\Gamma^{\rho\nu\mu}\chi_\rho$$

$$= -i\chi_{\mu,\nu}^{*T}\Gamma^0\Gamma^{\mu\nu\rho}\chi_\rho$$

$$= \partial_\nu(-i\chi_\mu^{*T}\Gamma^0\Gamma^{\mu\nu\rho}\chi_\rho)$$

$$+ i\chi_\mu^{*T}\Gamma^0\Gamma^{\mu\nu\rho}\chi_{\rho,\nu}$$

proving our claim provided that we replace D_ν by ∂_ν. If we take the connection into account, ie

$$D_\nu\chi_\rho = \partial_\nu\chi_\rho + (1/4)\omega_\nu^{mn}\Gamma_{mn}\chi_\rho$$

$$-\Gamma^\alpha_{\rho\nu}\chi_\alpha$$

then it follows that we must prove the skew-Hermitianity of the operator fields

$$\chi_\mu^{*T}\Gamma^0\Gamma^{\mu\nu\rho}\Gamma_{mn}\chi_\rho.\omega_\nu^{mn} --- (a)$$

and

$$\chi_\mu^{*T}\Gamma^0\Gamma^{\mu\nu\rho}\chi_\alpha.\Gamma^\alpha_{\rho\nu} --- (b)$$

However the field (b) is identically zero since $\Gamma^{\mu\nu\rho}$ is antisymmetric in (ν, ρ) while $\Gamma^\alpha_{\rho\nu}$ is symmetric in (ν, ρ). Hence, we have to prove only the skew-Hermitianity of the field

$$\chi_\mu^{*T}\Gamma^0\Gamma^{\mu\nu\rho}\Gamma_{mn}\chi_\rho --- (c)$$

Now,

$$\Gamma^{\mu\nu\rho}\Gamma_{mn} = (1/2)[\Gamma^{\mu\nu\rho}, \Gamma_{mn}] + (1/2)\{\Gamma_{mn}\Gamma^{\mu\nu\rho}\}$$

and

$$[\Gamma_{pqr}, \Gamma_{mn}] = [\Gamma_p\Gamma_{qr} + \Gamma_q\Gamma_{rp} + \Gamma_r\Gamma_{pq}, \Gamma_{mn}]$$

Now,

$$[\Gamma_p\Gamma_{qr}, \Gamma_{mn}] = \Gamma_p[\Gamma_{qr}, \Gamma_{mn}] + [\Gamma_p, \Gamma_{mn}]\Gamma_{qr}$$

$$= 4\Gamma_p(\eta_{qn}\Gamma_{rm} + \eta_{rm}\Gamma_{qn} - \eta_{qm}\Gamma_{rn} - \eta_{rn}\Gamma_{qm})$$
$$+4(\eta_{pm}\Gamma_n - \eta_{pn}\Gamma_m)\Gamma_{qr}$$

Summing this equation over cyclic permutations of (pqr) gives us

$$[\Gamma_{pqr}, \Gamma_{mn}] =$$
$$4\sum_{(pqr)} \eta_{mq}(\Gamma_p\Gamma_{nr} + \Gamma_r\Gamma_{pn} + \Gamma_n\Gamma_{rp})$$
$$+4\sum_{(pqr)} \eta_{nq}(\Gamma_p\Gamma_{rm} + \Gamma_r\Gamma_{mp} + \Gamma_m\Gamma_{pr})$$
$$= 4\sum_{(pqr)} (\eta_{mq}\Gamma_{pnr} + \eta_{nq}\Gamma_{prm})$$

Note that this quantity is antisymmetric w.r.t interchange of (m,n). It thus follows that

$$e_{q\nu}\chi_\mu^{*T}\Gamma^0[\Gamma^{\mu\nu\rho}, \Gamma_{mn}]\chi_\rho$$
$$= \chi^{p*T}\Gamma^0[\Gamma_{pqr}, \Gamma_{mn}]\chi^r$$
$$= 4\sum_{(pqr)} [\eta_{mq}\chi^{p*T}\Gamma^0\Gamma_{pnr}\chi^r + \eta_{nq}\chi^{p*T}\Gamma^0\Gamma_{prm}\chi^r]$$

Now,

$$(\chi^{p*T}\Gamma^0\Gamma_{pnr}\chi^r)^* =$$
$$\chi^{r*T}\Gamma^0\Gamma^{rnp}\chi^p$$
$$= \chi^{p*T}\Gamma^0\Gamma^{pnr}\chi^r$$

which proves the Hermitianity of $\chi^{p*}\Gamma^0\Gamma_{pnr}\chi^r$ and hence of $\bar{\chi}_\mu[\Gamma^{\mu\nu\rho}, \Gamma_{mn}]\chi_\rho$. In fact this quantity is identically zero. To see this, we use the Majorana Fermion property of χ^p to write

$$\chi^{p*T}\Gamma^0\Gamma_{pnr}\chi^r =$$
$$\chi^{pT}\Gamma^5\epsilon\Gamma_{pnr}\chi^r$$

and use the fact that

$$(\Gamma^5\epsilon\Gamma_{pnr})^T = -\Gamma_{pnr}^T\epsilon\Gamma^5$$
$$= -\epsilon\Gamma_{rnp}\Gamma^5 = \Gamma^5\epsilon\Gamma_{rnp} = -\Gamma^5\epsilon\Gamma_{pnr}$$

where we have used the identities

$$\Gamma_n^T\epsilon = \epsilon\Gamma_n, \Gamma_n\Gamma^5 = -\Gamma^5\Gamma_n$$

Then from the anticommutativity of the $\chi^{p's}$, we get

$$\chi^{pT}\Gamma^5\epsilon\Gamma_{pnr}\chi^r =$$
$$= -\chi^{rT}(\Gamma^5\epsilon\Gamma_{pnr})^T\chi^p =$$

$$= \chi^{rT}\Gamma^5\epsilon\Gamma_{pnr}\chi^p$$

$$= \chi^{pT}\Gamma^5\epsilon\Gamma_{rnp}\chi^r = -\chi^{pT}\Gamma^5\Gamma_{pnr}\chi^r$$

from which we conclude that

$$\chi^{pT}\Gamma^5\epsilon\Gamma_{rnp}\chi^r = 0$$

Now consider

$$X = \chi_\mu^{*T}\Gamma^0\{\Gamma^{\mu\nu\rho}, \Gamma_{mn}\}\chi_\rho$$

where $\{.,.\}$ denotes anticommutator. We have

$$X = X_1 + X_2$$

where

$$X_1 = \chi_\mu^{*T}\Gamma^0\Gamma^{\mu\nu\rho}\Gamma_{mn}\chi_\rho$$

$$X_2 = \chi_\mu^{*T}\Gamma^0\Gamma_{mn}\Gamma^{\mu\nu\rho}\chi_\rho$$

we have

$$X_1^* = \chi_\rho^{*T}\Gamma^0\Gamma_{mn}\Gamma^{\mu\nu\rho}\chi_\mu$$

$$= -\chi_\mu^{*T}\Gamma^0\Gamma_{mn}\Gamma^{\mu\nu\rho}\chi_\rho = -X_2$$

which shows that

$$X^* = -X$$

ie, X is skew Hermitian. Note that we have used the fact that

$$(\Gamma^0\Gamma_p\Gamma_q\Gamma_r\Gamma_m\Gamma_n)^* =$$

$$(\Gamma^0\Gamma_p\Gamma^0\Gamma^0\Gamma_q\Gamma_r\Gamma^0\Gamma^0\Gamma_m\Gamma^0\Gamma^0\Gamma_n)^* =$$

$$\Gamma^0\Gamma_n\Gamma_m\Gamma^0\Gamma^0\Gamma_r\Gamma^0\Gamma^0\Gamma_q\Gamma_p\Gamma^0\Gamma^0 =$$

$$\Gamma^0\Gamma_n\Gamma_m\Gamma_r\Gamma_q\Gamma_p$$

since $\Gamma^0\Gamma_n, \Gamma_n\Gamma^0, \Gamma^0$ are Hermitian and $\Gamma^{02} = I$. Thus, by antisymmetrizing over (pqr) and over (mn), we get

$$(\Gamma^0\Gamma_{pqr}\Gamma_{mn})^* = \Gamma^0\Gamma_{nm}\Gamma_{rqp}$$

$$= \Gamma^0\Gamma_{mn}\Gamma_{pqr}$$

This proves that

$$i\bar{\chi}_\mu\Gamma^{\mu\nu\rho}D_\nu\chi_\rho$$

is a Hermitian operator field.

Now consider the following local supersymmetry transformation

$$\delta\chi_\mu(x) = D_\mu\epsilon(x), \delta e_\mu^n = K\bar{\epsilon}(x)\Gamma^n\chi_\mu(x)$$

where $\epsilon(x)$ is an infinitesimal Majorana Fermionic parameter. We can easily check that $D_\mu\epsilon$ also satisfies the Majorana Fermion property. Indeed,

$$((D_\mu\epsilon(x))^*)^T = \partial_\mu(\epsilon(x)^*)^T + (\epsilon(x)^*)^T\Gamma^*_{mn}\omega^{mn}_\mu$$

$$= (\partial_\mu\epsilon(x))^T\Gamma^5\epsilon\Gamma^0 + \epsilon(x)^T\Gamma^5\epsilon\Gamma^0\Gamma^*_{mn}\omega^{mn}_\mu$$

$$= (\partial_\mu\epsilon(x))^T\Gamma^5\epsilon\Gamma^0 - \epsilon(x)^T\Gamma^5\epsilon\Gamma_{mn}\Gamma^0\omega^{mn}_\mu$$

Now,

$$\epsilon\Gamma_{mn} = -\Gamma^T_{mn}\epsilon$$

since

$$\Gamma^T_n\epsilon = \epsilon\Gamma_n$$

Thus, since $\Gamma^{0T} = \Gamma^0$, it follows that

$$\Gamma^5\epsilon\Gamma_{mn}\Gamma^0 = \Gamma^5\epsilon\Gamma_{mn}\Gamma^0 =$$

$$-\Gamma^5\Gamma^T_{mn}\epsilon\Gamma^0 = -\Gamma^T_{mn}\Gamma^5\epsilon\Gamma^0$$

This gives

$$((D_\mu\epsilon(x))^*)^T =$$

$$(\partial_\mu\epsilon(x))^T\Gamma^5\epsilon\Gamma^0$$

$$+\epsilon(x)^T\Gamma^T_{mn}\Gamma^5\epsilon\Gamma^0$$

$$= (D_\mu\epsilon(x))^T\Gamma^5\epsilon\Gamma^0$$

proving thereby the Majorana property of $D_\mu\epsilon(x)$. Now under the local supersymmetry transformation of χ_μ, the Gravitino Lagrangian changes by

$$\delta_\chi(\bar{\chi}_\mu\Gamma^{\mu\nu\rho}D_\nu\chi_\rho) =$$

$$= \delta\bar{\chi}_\mu\Gamma^{\mu\nu\rho}D_\nu\chi_\rho +$$

$$\bar{\chi}_\mu\Gamma^{\mu\nu\rho}D_\nu\delta\chi_\rho$$

$$= \bar{D}_\mu\epsilon(x)\Gamma^{\mu\nu\rho}D_\nu\chi_\rho$$

$$+\bar{\chi}_\mu\Gamma^{\mu\nu\rho}D_\nu D_\rho\epsilon(x)$$

The term in this quantity that is quadratic in $\{\omega^{mn}_\mu\}$ is given by

$$-\omega^{mn}_\mu\omega^{rs}_\rho[\epsilon(x)^T\Gamma^T_{mn}\Gamma^5\epsilon\Gamma^{\mu\nu\rho}\Gamma_{rs}\chi_\nu$$

$$+\chi^T_\nu\Gamma^5\epsilon\Gamma^{\mu\nu\rho}\Gamma_{mn}\Gamma_{rs}\epsilon(x)]$$

we must first prove that this is Hermitian under the assumption that $\epsilon(x)$ and $\chi_\mu(x)$ are mutually anticommuting Majorana Fermionic fields. Note that we can also express this as

$$-\omega^{mn}_\mu\omega^{rs}_\rho[\epsilon(x)^{*T}\Gamma^*_{mn}\Gamma^0\Gamma^{\mu\nu\rho}\Gamma_{rs}\chi_\nu(x)$$

$$+\chi_\nu(x)^{*T}\Gamma^0\Gamma^{\mu\nu\rho}\Gamma_{mn}\Gamma_{rs}\epsilon(x)]$$

Now,

$$\Gamma^*_{mn}\Gamma^0 = -\Gamma^0\Gamma_{mn}$$

Thus,

$$\Gamma^*_{mn}\Gamma^0\Gamma_{pqk}\Gamma_{rs} =$$

$$-\Gamma^0\Gamma_{mn}\Gamma_{pqk}\Gamma_{rs}$$

Thus,

$$(\epsilon(x)^{*T}\Gamma^*_{mn}\Gamma^0\Gamma^{\mu\nu\rho}\Gamma_{rs}\chi_\nu)^* =$$

$$-(\epsilon(x)^{*T}\Gamma^0\Gamma_{mn}\Gamma^{\mu\nu\rho}\Gamma_{rs}\chi_\nu)^* =$$

$$\chi^*_\nu\Gamma^0\Gamma_{rs}\Gamma^{\mu\nu\rho}\Gamma_{mn}\epsilon(x)$$

$$= \chi^T_\nu\Gamma^5\epsilon\Gamma_{rs}\Gamma^{\mu\nu\rho}\Gamma_{mn}\epsilon(x)$$

$$= -\epsilon(x)^T(\Gamma^5\epsilon\Gamma_{rs}\Gamma^{\mu\nu\rho}\Gamma_{mn})^T\chi_\nu$$

$$= \epsilon(x)^T\Gamma^T_{mn}(\Gamma^{\mu\nu\rho})^T\Gamma^T_{rs}\Gamma^5\epsilon\chi_\nu$$

$$= \epsilon(x)^T\Gamma^5\epsilon\Gamma_{mn}\Gamma^{\mu\nu\rho}\Gamma_{rs}\chi_\nu$$

$$= (\epsilon(x)^{*T}\Gamma^0\Gamma_{mn}\Gamma^{\mu\nu\rho}\Gamma_{rs}\chi_\nu$$

$$= -\epsilon(x)^{*T}\Gamma^*_{mn}\Gamma^0\Gamma^{\mu\nu\rho}\Gamma_{rs}\chi_\nu$$

This proves that $(\epsilon(x)^{*T}\Gamma^*_{mn}\Gamma^0\Gamma^{\mu\nu\rho}\Gamma_{rs}\chi_\nu)$ is skew-Hermitian. Note that if we replace $\epsilon(x)$ by $i\epsilon(x)$ where $\epsilon(x)$ is a Majorana Fermion, then the above quantity becomes Hermitian. Consider now the second term. It is

$$\chi^T_\nu\Gamma^5\epsilon\Gamma^{\mu\nu\rho}\Gamma_{mn}\Gamma_{rs}\epsilon(x)]$$

$$= \chi^*_\nu\Gamma^0\Gamma^{\mu\nu\rho}\Gamma_{mn}\Gamma_{rs}\epsilon(x)$$

$$= \chi^*_\nu\Gamma^0[\Gamma^{\mu\nu\rho},\Gamma_{mn}]\Gamma_{rs}\epsilon(x)$$

$$+\chi^*_\nu\Gamma^0\Gamma_{mn}\Gamma^{\mu\nu\rho}\Gamma_{rs}\epsilon(x)$$

This is also easily shown to be skew-Hermitian. Indeed, its adjoint is given by

[i] The general theory of Chiral superfields: The general superfield has the form

$$S(x,\theta) = C(x) + \theta^T\epsilon\omega(x) + \theta^T\epsilon\theta M(x) + \theta^T\gamma^5\epsilon N(x) + \theta^T\epsilon\gamma^\mu\theta V_\mu(x)$$

$$\theta^T\epsilon\theta.\theta^T\gamma^5\epsilon(\lambda(x) + a.\gamma^\nu\omega_{,\nu}(x) + (\theta^T\epsilon\theta)^2(D(x) + b.\Box C(x))$$

Under the above mentioned supersymmetry transformation

$$\alpha^T\gamma^5\epsilon.L = \bar{\alpha}L$$

where

$$L = \gamma^\mu\theta\partial_\mu + \gamma^5\epsilon\partial_\theta$$

the change in the superfield S is given by

$$\delta S = \bar{\alpha} L S$$

Here α is a Majorana Fermionic parameter. One can compute the change in the component fields $C, \omega, M, N, V_\mu, \lambda, D$ under this infinitesimal supersymmetry transformation and show that D changes by a perfect space-time four divergence and hence can be used as a candidate Lagrangian. However, in the case when the superfield is such that $\lambda = 0, D = 0, V_\mu = B_{,\mu}$, the resulting superfield is called Chiral and it is easy to prove that under a supersymmetry transformation, a Chiral superfield transforms into a Chiral superfield. The general Chiral superfield can be expressed as

$$S_c(x,\theta) = C(x) + \theta^T \epsilon\omega(x) + \theta^T \epsilon\theta M(x) + \theta^T \gamma^5 \epsilon\theta N(x) + \theta^T \epsilon\gamma^\mu \theta B_{,\mu}(x)$$
$$+ a\theta^T \epsilon\theta.\theta^T \gamma^5 \epsilon\gamma^\nu \omega_{,\nu}(x) + b(\theta^T \epsilon\theta)^2 \Box C(x) --- (1)$$

Let us first prove that the class of Chiral supefields is supersymmetry invariant. To do so, we first note that for the superfield S_c above, the change in λ under the above infinitesimal supersymmetry transformation, obtained by equating the cubic terms in θ, is given by

$$\theta^T \epsilon\theta.\theta^T \gamma^5 \epsilon(\delta\lambda + \gamma^\nu \delta\omega_{,\nu})$$
$$= \bar{\alpha}\gamma^\mu \theta(\theta^T \epsilon\theta.M_{,\mu} + \theta^T \gamma^5 \epsilon\theta N_{,\mu} + \theta^T \epsilon\gamma^\nu \theta.B_{,\nu\mu})$$
$$+ \bar{\alpha}\gamma^5 \epsilon(4b.\theta^T \epsilon\theta.\epsilon\theta)\Box C$$

The change in ω is given by equating linear terms in θ:

$$\theta^T \epsilon.\delta\omega = \bar{\alpha}\gamma^\mu \theta C_{,\mu} +$$
$$\bar{\alpha}\gamma^5 \epsilon(2\epsilon\theta.M + 2\gamma^5 \epsilon N + 2\epsilon\gamma^\nu \theta.B_{,\nu})$$

By using the identities

$$\epsilon\gamma^\mu, \epsilon, \gamma^5 \epsilon$$

are skewsymmetric,

$$\theta\theta^T = (1/4)(\theta^T \epsilon\theta.\epsilon + \theta^T \gamma^5 \epsilon\gamma^5 \epsilon + \theta^T \epsilon\gamma^\mu \theta.\epsilon\gamma_\mu) --- (a)$$

and

$$\{\gamma^\mu, \gamma^\nu\} = 2\eta^{\mu\nu}$$

it is easy to see from the above relations that

$$\delta\lambda = 0$$

Likewise we can verify that $\delta D = 0$ and the condition that $V_\mu = B_{,\mu}$, ie, V_μ is a perfect four gradient also remains invariant under a supersymmetry transformation. Indeed,

$$(\theta^T \epsilon\theta)^2 (\delta D + b.\Box\delta C)$$
$$= \bar{\alpha}\gamma^\mu \theta(\theta^T \epsilon\theta).\theta^T \gamma^5 \epsilon(a.\gamma^\nu \omega_{,\mu\nu})$$

and

$$\delta C = \bar{\alpha}\gamma^5 \epsilon \partial_\theta \theta^T \epsilon \omega$$
$$= \alpha^T \epsilon \omega$$

together imply that

$$\delta D = 0$$

Note that the product of any two distinct members of the six quantities $\theta^T \epsilon \theta, \theta^T \gamma^5 \epsilon \theta, \theta^T \epsilon \gamma^\mu \theta$ is zero and therefore

$$\theta(\theta^T \epsilon \theta)\theta^T = (\theta^T \epsilon \theta)\epsilon/4$$

We also use

$$\gamma^\mu \gamma^\nu \omega_{,\mu\nu} = \eta^{\mu\nu} \omega_{,\mu\nu} = \Box \omega$$

Finally, we must verify that $V_\mu = B_{,\mu}$ changes by an exact four gradient under a supersymmetry transformation. To see this we use the equations corresponding to quadratic terms in the θ:

$$\theta^T \epsilon \theta . \delta M + \theta^T \gamma^5 \epsilon \theta . \delta N + \theta^T \epsilon \gamma^\mu \theta \delta V_\mu =$$
$$\bar{\alpha} \gamma^\mu \theta \theta^T \epsilon \omega_{,\mu} + \alpha^T \partial_\theta (\theta^T \epsilon \theta . \theta^T \gamma^5 \epsilon (a . \gamma^\mu \omega_{,\mu})$$
$$= \alpha^T \gamma^5 \epsilon \gamma^\mu \theta \theta^T \epsilon \omega_{,\mu} +$$
$$\alpha^T (2\epsilon \theta \theta^T + \theta^T \epsilon \theta \gamma^5 \epsilon)(a \gamma^\mu \omega_{,\mu})$$

Noting that the terms $\theta^T \epsilon \theta, \theta^T \gamma^5 \epsilon \theta, \theta^T \epsilon \gamma^\mu \theta$ are all the six linearly independent quadratic combinations of the θ, we get on equating the coefficients of $\theta^T \epsilon \gamma^\mu \theta$ on both sides after recalling identity (a) that

$$\delta V_\mu = \alpha^T \gamma^5 \epsilon \gamma^\nu (1/4) \epsilon \gamma_\mu \omega_{,\nu}$$
$$+ (a/2) \alpha^T \epsilon . \epsilon \gamma_\mu \gamma^\nu \omega_{,\nu}$$

which indeed proves that δV_μ is a perfect four gradient thereby completing the proof that the class of Chiral fields is closed under supersymmetry transformations.

Remark: Matter fields in non-Abelian quantum field theory get generalized in supersymmetry theory to super matter fields which are obtained from the D-component of products of left Chiral fields with their complex conjugates while gauge fields in quantum field theory get generalized to super-gauge fields which are derived from hte gauge fields V_μ, the gaugino fields λ and the auxiliary fields D. In conventional non-Abelian quantum field theory, the matter fields transform according to a representation of the gauge group with the group element being in general local, ie, a function of the space-time coordinates while the gauge fields transform according to the adjoint representation of the gauge group plus an additional factor involving space-time gradients of the representation of the local gauge group elements. In supersymmetry theory, the matter field Lagrangian derived from D-component of quadratic combinations of left Chiral fields and their complex conjugates comprises of the scalar field part, the Dirac spinor field part and and auxiliary part which is determined in terms of the first two parts by setting the variational derivative of the corresponding action w.r.t it to zero, ie, it is determined by its field equation which is a purely algebraic equation for it. Supersymmetry predicts then that the mass term in the Dirac field component Lagrangian contains a mass term that depends on the scalar field.

Thus an arbitrary Chiral superfield has the form (1) which we repeat here for convenience

$S_c(x,\theta) = C(x)+\theta^T\epsilon\omega(x)+$

$\theta^T\epsilon\theta M(x)+\theta^T\gamma^5\epsilon\theta N(x)+\theta^T\epsilon\gamma^\mu\theta B_{,\mu}(x)+a\theta^T\epsilon\theta.\theta^T\gamma^5\epsilon\gamma^\nu\omega_{,\nu}(x)+b(\theta^T\epsilon\theta)^2\Box C(x)---(1)$

Now,

$$\theta^T\epsilon\theta = \theta_R^T\epsilon\theta_R + \theta_L^T\epsilon\theta_L$$

$$\theta^T\gamma^5\epsilon\theta = \theta_R^T\gamma^5\epsilon\theta_R + \theta_L^T\gamma^5\epsilon\theta_L$$

$$= -\theta_R^T\epsilon\theta_R + \theta_L^T\epsilon\theta_L$$

since

$$\theta_R^T\epsilon\theta_L = 0, \theta_R^T\gamma^5\epsilon\theta_L = 0$$

and

$$\theta = \theta_R + \theta_L = (1-\gamma^5)\theta/2 + (1+\gamma^5)\theta/2$$

Also since

$$\theta_R^T\epsilon\gamma^\mu\theta_R = \theta_L^T\epsilon\gamma^\mu\theta_L = 0$$

it follows that

$$\theta^T\epsilon\gamma^\mu\theta = 2\theta_R^T\epsilon\gamma^\mu\theta_L$$

$$= 2\theta_L^T\epsilon\gamma^\mu\theta_R$$

since θ_R and θ_L anticommute and $\epsilon\gamma^\mu$ is skew-symmetric. Further,

$$\theta^T\epsilon\theta.\theta^T\gamma^5\epsilon =$$

$$(\theta_R^T\epsilon\theta_R + \theta_L^T\epsilon\theta_L)(\theta_R^T + \theta_L^T)\gamma^5\epsilon$$

$$= \theta_R^T\epsilon\theta_R.\theta_L^T\gamma^5\epsilon + \theta_L^T\epsilon\theta_L\theta_R^T\gamma^5\epsilon$$

Now observe that

$$\theta_R\theta_L^T = (1/4)(1-\gamma^5)\theta\theta^T(1+\gamma^5)$$

$$= (1/16)(1-\gamma^5)[\theta^T\epsilon\theta\epsilon + \theta^T\gamma^5\epsilon\theta\gamma^5\epsilon + \theta^T\epsilon\gamma^\mu\theta\epsilon\gamma_\mu](1+\gamma^5)$$

$$= (1/8)(1-\gamma^5)\epsilon\gamma_\mu(1+\gamma^5)\theta_R^T\epsilon\gamma^\mu\theta_L$$

$$= (1/4)(1-\gamma^5)\epsilon\gamma_\mu\theta_R^T\epsilon\gamma^\mu\theta_L$$

and likewise,

$$\theta_L.\theta_R^T = (1/4)(1+\gamma^5)\epsilon\gamma_\mu\theta_R^T\epsilon\gamma^\mu\theta_L$$

Thus, using the fact that

$$\theta_R^T(1-\gamma^5)/2 = \theta_R^T, \theta_L^T(1+\gamma^5)/2 = \theta_L$$

we get

$$\theta^T\epsilon\theta.\theta^T =$$

$$\theta_R^T\epsilon\theta_R^T\epsilon\gamma^\mu\theta_L\epsilon\gamma_\mu/2$$

$$+\theta_L^T\epsilon\theta_L^T\epsilon\gamma^\mu\theta_R\epsilon\gamma_\mu/2$$

$$= -\theta_R^T\epsilon\gamma^\mu\theta_L\theta_R^T\gamma_\mu/2$$

$$-\theta_R^T\epsilon\gamma^\mu\theta_L\theta_L^T\gamma_\mu/2$$

Another way to see this is as follows:

$$\theta^T\epsilon\theta\theta^T =$$

$$(\theta_R^T + \theta_L^T)\epsilon(\theta^T\epsilon\theta\epsilon + \theta^T\gamma^5\epsilon\theta\gamma^5\epsilon + \theta^T\epsilon\gamma^\mu\theta.\epsilon\gamma_\mu)/4$$

$$= (\theta_R^T + \theta_L^T)\epsilon(\theta_R^T\epsilon\theta_R\epsilon + \theta_L^T\epsilon\theta_L\epsilon + \theta_R^T\gamma^5\epsilon\theta_R\gamma^5\epsilon + \theta_L^T\gamma^5\epsilon\theta_L\gamma^5\epsilon + 2\theta_R^T\epsilon\gamma^\mu\theta_L.\epsilon\gamma_\mu)/4$$

Now,

$$\theta_R^T\gamma^5\epsilon\theta_R = -\theta_R^T\epsilon\theta_R,$$

$$\theta_L^T\gamma^5\epsilon\theta_L = \theta_L^T\epsilon\theta_L$$

and hence

$$\theta_R^T\epsilon(\theta_L^T\epsilon\theta_L\epsilon + \theta_L^T\gamma^5\epsilon\theta_L\gamma^5\epsilon)$$

$$= \theta_R^T\epsilon(\theta_L^T\epsilon\theta_L(1+\gamma^5)\epsilon/2$$

$$= -\theta_R^T(1+\gamma^5)\theta_L^T\epsilon\theta_L/2 = 0$$

and likewise,

$$\theta_L^T\epsilon(\theta_R^T\epsilon\theta_R\epsilon + \theta_R^T\gamma^5\epsilon\theta_R\gamma^5\epsilon)$$

$$= -\theta_L^T\theta_R^T\epsilon\theta_R(1-\gamma^5)/2 = 0$$

Thus, we get

$$\theta^T\epsilon\theta\theta^T = -(1/2)\theta_R^T\gamma_\mu\theta_R^T\epsilon\gamma^\mu\theta_L - (1/2)\theta_L^T\gamma_\mu\theta_R^T\epsilon\gamma^\mu\theta_L$$

Thus, the Chiral superfield (1) can be expressed as

$$S_c(x,\theta) =$$

$$C(x) + (\theta_L^T\epsilon\omega(x) + \theta_R^T\epsilon\omega(x)) +$$

$$(\theta_L^T\epsilon\theta_L + \theta_R^T\epsilon\theta_R)M(x)$$

$$+(\theta_L^T\epsilon\theta_L - \theta_R^T\epsilon\theta_R)N(x)$$

$$+2\theta_R^T\epsilon\gamma^\mu\theta_L B_{,\mu}(x)$$

$$-(a/2)(\theta_R^T\epsilon\gamma^\mu\theta_L)(\theta_L^T\gamma_\mu\gamma^5\epsilon\gamma^\nu\omega_{,\nu} + \theta_R^T\gamma_\mu\gamma^5\epsilon\gamma^\nu\omega_{,\nu})$$

$$+(1/2)(\theta_R^T\epsilon\gamma^\mu\theta_L).(\theta_R^T\epsilon\gamma_\mu\theta_L)b\Box C$$

An alternate more convenient formula for the cubic term is obtained as follows: We've already noted that

$$\theta^T\epsilon\theta.\theta^T =$$

$$(\theta_R^T\epsilon\theta_R + \theta_L^T\epsilon\theta_L)(\theta_R^T + \theta_L^T)$$

$$= \theta_R^T\epsilon\theta_R.\theta_L^T + \theta_L^T\epsilon\theta_L.\theta_R^T$$

On the other hand, consider the expression

$$A = \theta_R^T\epsilon\gamma^\mu\theta_L.\theta_L^T\epsilon.\omega_{,\mu}$$

We have

$$\theta_L\theta_L^T = ((1+\gamma^5)/2)\theta\theta^T(1+\gamma^5)/2$$

$$= ((1+\gamma^5)/2)(\theta^T\epsilon\theta\epsilon + \theta^T\gamma^5\epsilon\theta\gamma^5\epsilon)((1+\gamma^5)/2)(1/4)$$

$$= \theta^T(1+\gamma^5)\epsilon\theta((1+\gamma^5)/8)$$

$$= \theta_L^T\epsilon\theta_L(1+\gamma^5)\epsilon/4$$

Thus,

$$A = (1/4)\theta_L^T\epsilon\theta_L.\theta_R^T\epsilon\gamma^\mu(1+\gamma^5)\epsilon\epsilon\omega_{,\nu}$$

$$= -(1/2)\theta_L^T\epsilon\theta_L.\theta_R^T\epsilon\gamma^\mu\omega_{,\mu}$$

$$= (1/2)\theta_L^T\epsilon\theta_L.\theta_R^T\gamma^5\epsilon\gamma^\mu\omega_{,\mu}$$

Likewise, defining

$$B = \theta_R^T\epsilon\gamma^\mu\theta_L.\theta_R^T\epsilon.\omega_{,\mu}$$

$$= \theta_L^T\epsilon\gamma^\mu\theta_R.\theta_R^T\epsilon\omega_{,\mu}$$

we get using

$$\theta_R\theta_R^T = ((1-\gamma^5)/2)\theta\theta^T((1+\gamma^5)/2)$$

$$= ((1-\gamma^5)/2)(\theta^T\epsilon\theta\epsilon + \theta^T\gamma^5\epsilon\theta\gamma^5\epsilon)((1-\gamma^5)/2)(1/4)$$

$$= ((1-\gamma^5)/8)\epsilon.\theta^T(1-\gamma^5)\epsilon\theta$$

$$= ((1-\gamma^5)/4)\epsilon.\theta_R^T\epsilon\theta_R$$

and hence,

$$B = -\theta_L^T\epsilon\gamma^\mu((1-\gamma^5)/4)\omega_{,\mu}\theta_R^T\epsilon\theta_R$$

$$= (-1/2)\theta_R^T\epsilon\theta_R.\theta_L^T\gamma^5\epsilon\gamma^\mu\omega_{,\mu}$$

Thus, the cubic term in the above Chiral field is given by

$$a\theta^T\epsilon\theta.\theta^T\gamma^5\epsilon\gamma^\mu\omega_{,\mu}$$

$$= a(\theta_L^T\epsilon\theta_L\theta_R^T\gamma^5\epsilon\gamma^\mu\omega_{,\mu} + \theta_R^T\epsilon\theta_R\theta_L^T\gamma^5\epsilon\gamma^\mu\omega_{,\mu})$$

$$= 2a(\theta_R^T\epsilon\gamma^\mu\theta_L\theta_L^T\epsilon\omega_{,\mu}$$

$$-\theta_R^T\epsilon\gamma^\mu\theta_L\theta_R^T\epsilon\omega_{,\mu})$$

Combining these two identities, we can express the general Chiral super-field as

$$S_c(x,\theta) =$$

$$C(x) + (\theta_L^T \epsilon\omega(x) + \theta_R^T \epsilon\omega(x)) +$$

$$(\theta_L^T \epsilon\theta_L + \theta_R^T \epsilon\theta_R)M(x)$$

$$+(\theta_L^T \epsilon\theta_L - \theta_R^T \epsilon\theta_R)N(x)$$

$$+2\theta_R^T \epsilon\gamma^\mu \theta_L B_{,\mu}(x)$$

$$+2a\theta_R^T \epsilon\gamma^\mu \theta_L (\theta_L^T \epsilon\omega_{,\mu} - \theta_R^T \epsilon\omega_{,\mu})$$

$$+(1/2)(\theta_R^T \epsilon\gamma^\mu \theta_L).(\theta_R^T \epsilon\gamma_\mu \theta_L) b\Box C$$

Note:

$$(\theta_R^T \epsilon\gamma^\mu \theta_L).(\theta_R^T \epsilon\gamma_\mu \theta_L)$$

$$= \theta_R^T \epsilon\gamma^\mu \theta_L \theta_L^T \epsilon\gamma_\mu \theta_R$$

and

$$\theta_L \theta_L^T = (1/4)(1+\gamma^5)\theta.\theta^T(1+\gamma^5)$$

$$= (1/4)(\theta^T \epsilon\theta\epsilon(1+\gamma^5)^2 + \theta^T \gamma^5 \epsilon\theta\gamma^5 \epsilon(1+\gamma^5)^2)$$

$$= (1/2)\theta^T \epsilon(1+\gamma^5)\theta.(1+\gamma^5)\epsilon$$

$$= \theta_L^T \epsilon\theta_L (1+\gamma^5)\epsilon$$

and hence,

$$(\theta_R^T \epsilon\gamma^\mu \theta_L).(\theta_R^T \epsilon\gamma_\mu \theta_L)$$

$$= (\theta_R^T \epsilon\gamma^\mu (1+\gamma^5)\epsilon.\epsilon\gamma_\mu \theta_R).(\theta_L^T \epsilon\theta_L)$$

$$= 4(\theta_R^T \epsilon\theta_R).(\theta_L^T \epsilon\theta_L)$$

since γ^5 anticommutes with γ^μ and $\gamma^\mu \gamma_\mu = -2$. On the other hand,

$$(\theta^T \epsilon\theta)^2 = (\theta_L^T \epsilon\theta_L + \theta_R^T \epsilon\theta_R)^2$$

$$= 2(\theta_R^T \epsilon\theta_R).(\theta_L^T \epsilon\theta_L)$$

Exercise: Show that the general Chiral superfield $S_c(x,\theta)$ is expressible as the sum of a left Chiral superfield and a right Chiral superfield, where a left Chiral field is a function of θ_L and $x_+^\mu = x^\mu + \theta_R^T \epsilon\gamma^\mu \theta_L$ and conversely a right Chiral super-field is a function of θ_R and $x_-^{mu} x^\mu - \theta_R^T \epsilon\gamma^\mu \theta_R$. Specifically, any left Chiral super-field can be expressed as

$$\Phi(x,\theta) = \phi(x_+) + \theta_L^T \epsilon\psi(x_+) + \theta_L^T \epsilon\theta_L F(x)$$

where ψ is a left Chiral field and any right Chiral field can be expressed as

$$\eta(x,\theta) = \phi(x_-) + \theta_R^T\epsilon\psi(x_-) + \theta_R^T\epsilon\theta_R F(x)$$

where ψ is right Chiral. Note that there is no loss of generality in assuming that ψ is left Chiral in the former case and right Chiral in the latter since if ψ is arbitrary, then

$$\theta_L^T\epsilon\psi = \theta_L^T\epsilon((1+\gamma^5)/2)\psi$$

$$\theta_R^T\epsilon\psi = \theta_R^T\epsilon((1-\gamma^5)/2)\psi$$

since γ^5 commutes with ϵ. Note that

$$\theta_L^* = \gamma^5\epsilon\gamma^0\theta_R,$$

$$\theta_R^* = \gamma^5\epsilon\gamma^0\theta_L$$

It is easy to verify that

$$\gamma^5\epsilon\gamma^0$$

is a real matrix whose square is the identity and this confirms the requirement that

$$(\theta_L^*)^* = \theta_L, (\theta_R^*)^* = \theta_R$$

Note that by the definition of the Majorana Fermion,

$$\theta^* = \gamma^5\epsilon\gamma^0\theta$$

which gives

$$(\theta_L + \theta_R)^* = \theta_L^* + \theta_R^* = \gamma^5\epsilon\gamma^0(\theta_L + \theta_R)$$

from which the desired relationships follow on equating the first two components and the last two components. Now,

$$(\theta_R^T\epsilon\gamma^\mu\theta_L)^*$$

$$= (\theta_R^*)^T conj(\epsilon\gamma^\mu)\theta_L^*$$

$$= (\gamma^5\epsilon\gamma^0\theta_L)^T conj(\epsilon\gamma^\mu)\gamma^5\epsilon\gamma^0\theta_R$$

$$= \theta_L^T\gamma^5\epsilon\gamma^0.conj(\epsilon\gamma^\mu)\gamma^5\epsilon\gamma^0\theta_R$$

$$= -\theta_L^T\gamma^5\epsilon\gamma^0(\epsilon\gamma^\mu)^*\gamma^5\epsilon\gamma^0\theta_R$$

since $\epsilon\gamma^\mu$ is skew-symmetric. Now since $\gamma^5\epsilon\gamma^0$ is Hermitian and γ^μ and γ^5 anticommute (note that γ^5 and ϵ commute), it follows that

$$\gamma^5\epsilon\gamma^0(\epsilon\gamma^\mu)^*\gamma^5\epsilon\gamma^0$$

$$= (\gamma^5\epsilon\gamma^0\epsilon\gamma^\mu\gamma^5\epsilon\gamma^0)^*$$

$$(\epsilon\gamma^0\epsilon\gamma^\mu\epsilon\gamma^0)^*$$

Now,

$$\epsilon\gamma^0 = \gamma^0\epsilon$$

and further,

$$(\epsilon\gamma^0)^* = -\gamma^0\epsilon$$

and $\gamma^0\gamma^\mu$ is Hermitian. Thus, the above equals

$$-(\gamma^0\gamma^\mu\epsilon\gamma^0)^*$$

$$= \gamma^0\epsilon\gamma^0\gamma^\mu = \epsilon\gamma^\mu$$

Thus, we have proved that

$$(\theta_R^T\epsilon\gamma^\mu\theta_L)^* = -\theta_L^T\epsilon\gamma^\mu\theta_R$$

$$= -\theta_R^T\epsilon\gamma^\mu\theta_L$$

It follows that

$$(x_+^\mu)^* = x_-^\mu$$

and since

$$\theta_L^* = \gamma^5\epsilon\gamma^0\theta_R$$

with $\gamma^5\epsilon\gamma^0$ being non-singular, the conjugate of any function of (θ_L, x_+^μ) is a function of (θ_R, x_-^μ) and conversely. Therefore, the conjugate of any left Chiral superfield is a right Chiral superfield and conversely.

[j] Construction of supersymmetric matter Lagrangians (actions) from Chiral superfields. We shall observe that when we construct the supersymmetric Lagrangian as $[\Phi^*\Phi]_D$, then we get terms corresponding to the kinetic parts of the Klein-Gordon and Dirac Lagrangians while when we construct the supersymmetric Lagrangian as $[f(\Phi)]_F$, we get extra inertial parts for these Lagrqangians. In particular, we will observe the remarkable fact that when after constructing the total Lagrangian by adding these two components, we write down the field equations for the auxiliary fields D, F, we will be able to eliminate these terms and will obtain a broken supersymmetric Lagrangian. In particular, the solution to the auxiliary field equations determine the masses of the Dirac particle. In this way, supersymmetry is able to explain in a natural way how the scalar field couples to the Dirac field giving rise to massive Dirac particles after supersymmetry is broken.

Consider now the Lagrangian

$$L_1 = [\Phi^*\Phi]_D$$

where

$$\Phi = \phi(x_+) + \theta_L^T\epsilon\psi(x_+) + \theta_L^T\epsilon\theta_L F(x)$$

Then,

$$\Phi^* = \phi^*(x_-) + \theta_R^T\gamma^5\epsilon\gamma^0\epsilon\psi^*(x_-) + \theta_R^T\gamma^5\epsilon\gamma^0\epsilon\gamma^5\epsilon\gamma^0\theta_R F^*(x)$$

$$= \phi^*(x_-) + \theta_R^T\gamma^0\psi^*(x_)+\theta_R^T\epsilon\theta_R F^*(x)$$

We shall now evaluate all the terms in $[\Phi^*\Phi]_D$: First,

$$[\phi^*(x_-)\phi(x_+)]_D = T_1 + T_2 + T_3$$

where

$$T_1 = -\phi^*_{,\mu}\phi_{,\nu}(\theta_R^T\epsilon\gamma^\mu\theta_L).(\theta_R^T\epsilon\gamma^\nu\theta_L]_D$$
$$= c_1\eta^{\mu\nu}\phi^*_{,\mu}(x)\phi_{,\nu}(x) = c_1\partial_\mu\phi^*(x).\partial^\mu\phi(x)$$
$$T_2 = \phi^*(x)\phi_{,\mu\nu}(x)[\theta_R^T\epsilon\gamma^\mu\theta_L.\theta_R^T\epsilon\gamma^\nu\theta_L]_D$$
$$= c_1\phi^*(x)\Box\phi(x)$$

where

$$\Box = \eta^{\mu\nu}\partial_\mu\partial_\nu = \partial^\mu\partial_\mu$$

and likewise,

$$T_3 = c_1\phi(x)\Box\phi^*(x)$$

Remark:

$$\theta_R^T\epsilon\gamma^\mu\theta_L.\theta_R^T\epsilon\gamma^\nu\theta_L$$
$$= \theta_R^T\epsilon\gamma^\mu\theta_L\theta_L^T\epsilon\gamma^\nu\theta_R$$
$$\theta_L\theta_L^T = ((1+\gamma^5)/2)\theta\theta^T((1+\gamma^5)/2)$$
$$= \theta\epsilon(1+\gamma^5)\theta((1+\gamma^5)/2)\epsilon$$
$$= \theta_L^T\epsilon\theta_L(1+\gamma^5)\epsilon$$

Thus,

$$\theta_R^T\epsilon\gamma^\mu\theta_L.\theta_R^T\epsilon\gamma^\nu\theta_L =$$
$$(\theta_L^T\epsilon\theta_L)\theta_R^T\epsilon\gamma^\mu\epsilon(1+\gamma^5)\epsilon\epsilon\gamma^\nu\theta_R$$
$$= -2(\theta_L^T\epsilon\theta_L).\theta_R^T\epsilon\gamma^\mu\epsilon\gamma^\nu\theta_R$$
$$= c_1(\theta^T\epsilon\theta)^2\eta^{\mu\nu}$$

which follows on expressing γ^μ in terms of the Pauli spin matrices and using the fact that θ_L has components $\theta_{1:2}$ while θ_R has components $\theta_{3:4}$ and the product of any four of the $\theta's$ is non-zero iff all the $\theta's$ are distinct. Again,

$$[\theta_R^T\gamma^0\psi^*(x_-)\theta_L^T\epsilon\psi(x_+)]_D$$
$$= [\psi^*_{,\mu}(x)^T\theta_R^T\epsilon\gamma^\mu\theta_L\gamma^0\theta_R\theta_L^T\epsilon\psi(x)]_D$$
$$+[\psi_{,\mu}(x)^T\theta_R^T\epsilon\gamma^\mu\theta_L\epsilon\theta_L\theta_R^T\gamma^0\psi^*(x)]_D$$

Now,

$$\theta_R\theta_L^T = \theta_R^T\epsilon\gamma^\mu\theta_L\epsilon\gamma_\mu(1+\gamma^5)$$

So

$$[\psi^*_{,\mu}(x)^T\theta_R^T\epsilon\gamma^\mu\theta_L\gamma^0\theta_R\theta_L^T\epsilon\psi(x)]_D$$
$$= 2[\theta_R^T\epsilon\gamma^\mu\theta_L.\theta_R^T\epsilon\gamma^\nu\theta_L]_D.\psi^*_{,\mu}(x)^T\gamma^0\epsilon\gamma_\nu\epsilon\psi(x)$$

$$= -2c_1\eta^{\mu\nu}\psi^*_{,\mu}(x)^T\gamma^0\gamma_\nu^T\psi(x)$$

$$= -2c_1\eta^{\mu\nu}\bar{\psi}_{,\mu}(x)\gamma_\nu^T\psi(x)$$

$$= -2c_1\eta^{\mu\nu}\psi^*_{,\mu}(x)^T\gamma_\nu^{*T}\gamma^0\psi(x)$$

$$= -2c_1(\gamma^\mu\psi_{,\mu}(x))^*\gamma^0\psi(x)$$

where we have used the fact that $\gamma_\mu\gamma^0$ is Hermitian and hence its transpose $\gamma^0\gamma_\mu^T$ is also Hermitian. Note that the conjugate of this quantity is given by

$$-2c_1\psi(x)^*\gamma^0\gamma^\mu\psi_{,\mu}(x) = -2c_1\bar{\psi}(x)\gamma^\mu\psi_{,\mu}(x)$$

$$= -2c_1\psi(x)^*\alpha^\mu\psi_{,\mu}(x)$$

[k] Feynnman superpath integrals and superpropagators.
The action functional for the left Chiral superfield Φ is taken as

$$S[\Phi] = \int [\Phi^*\Phi]_D d^4x + \int [f(\Phi)]_F d^4x$$

Equivalently, writing $\Phi = D_R^2 S$ we can express this as

$$S = \int (D_L^2 S^*)(D_R^2 S)d^4x d^4\theta + \int \tilde{S} d^4x d^4\theta$$

Note that

$$f(D_R^2 S) = D_R^2\tilde{S}$$

for some other superfield $\tilde{S}$. For example,

$$(D_R^2 S)^2 = D_R^2(SD_R^2 S), (D_R^2 S)^3 = D_R^2(S(D_R^2 S)^2)$$

etc. So in fact writing

$$f(\Phi) = \sum_{k\geq 1} c(k)\Phi^k$$

we get

$$f(D_R^2 S) = D_R^2(\sum_{k\geq 1} c(k)S(D_R^2 S)^{k-1})$$

So formally, we can write

$$f(D_R^2 S) = D_R^2(S.(D_R^2 S)^{-1}f(D_R^2 S))$$

The superfield equations are expressible as

$$D_R^2 D_L^2 S^* = f'(\Phi) = f'(D_R^2 S)$$

or equivalently,

$$D_L^2 D_R^2 S = f*'(D_L^2 S^*)$$

since

$$f'(D_R^2 S)D_R^2\delta S = D_R^2(f'(D_R^2 S)\delta S)$$

and hence

$$\int [f'(D_R^2 S)D_R^2\delta S]_F d^4x = \int f'(D_R^2 S)\delta S d^4x d^4\theta$$

By analogy with quantum field theory, it is therefore natural to consider a super-Green's function $G(x,\theta|x',\theta')$ that satisfies the propagator equation

$$D_L^2 D_R^2 G(x,\theta|x',\theta') - f'*(D_L^2 G(x,\theta|x',\theta')^*) = \delta^4(x-x')\delta^4(\theta-\theta')$$

Note that by the definition of the Berezin/Fermionic integral,

$$\int \theta_{i_1}..\theta_{i_k} d^4\theta$$

is zero if either $k < 4$ or if $k > 4$ while

$$\int \theta_1\theta_2\theta_3\theta_4 d^4\theta = 1$$

it follows that

$$\delta^4(\theta-\theta') = (\theta_1-\theta_1')(\theta_2-\theta_2')(\theta_3-\theta_3')(\theta_4-\theta_4')$$

This may be explicitly checked by writing out $f(\theta)$ as

$$f(\theta) = c_0 + c_1(k)\theta_k + c_2(k,m)\theta_k\theta_m + c_3(k,m,n)\theta_k\theta_m\theta_n + c_4\theta_1\theta_2\theta_3\theta_4$$

and applying the above Berezin rules taking into account the anitcommutativity of the θ and the θ' to show that

$$\int f(\theta)\delta^4(\theta-\theta')d^4\theta = f(\theta')$$

In the absence of a superpotential $f(\Phi)$, the field equations are

$$D_L^2 D_R^2 S = 0$$

This equation should be regarded as the super-version of the classical massless Klein-Gordon equation or equivalently the wave equation. The corresponding super-propagator G should satisfy the super pde

$$D_L^2 D_R^2 G = P\delta^4(x-x')\delta^4(\theta-\theta')$$

where P is the projection onto the space of superfields fields that belong to the orthogonal complement of the nullspace of $D_L^2 D_R^2$, or equivalently that belong to the range space of $D_L^2 D_R^2$. This is precisely in analogy with quantum electrodynamics. It is easy to see that

$$P = K\Box^{-1}D_L^2 D_R^2$$

for some real constant K. In fact, we have that P annihilates any vector in the range of D_R and hence in the range of D_R^2 and further, we have

$$P^2 = K^2\Box^{-2}D_L^2D_R^2D_L^2D_R^2 = D_L^2[D_R^2, D_L^2]D_R^2$$

with

$$D_R^2 D_{La} = D_R^T\epsilon D_R D_{La}$$

$$\{D_{Ra}, D_{Lb}\} = \{(\gamma^\mu\theta_L\partial_\mu - \gamma^5\epsilon\partial_{\theta_R})_a, (\gamma^\nu\theta_R\partial_\nu - \gamma^5\epsilon\partial_{\theta_L})_b\}$$
$$= -(\gamma^\mu)_{ac}\{\theta_{Lc}, \partial_{\theta_{Ld}}\}(\gamma^5\epsilon)_{bd}\partial_\mu$$
$$-(\gamma^5\epsilon)_{ac}\{\partial_{\theta_{Rc}}, \theta_{Rd}\}\gamma^\nu_{bd}\partial_\nu$$
$$= -[\gamma^\mu((1+\gamma^5)/2)(\gamma^5\epsilon)^T - \gamma^5\epsilon((1-\gamma^5)/2)\gamma^{\mu T}]_{ab}\partial_\mu$$
$$= [\gamma^\mu(1+\gamma^5)\epsilon + (1-\gamma^5)\gamma^\mu\epsilon]_{ab}\partial_\mu$$
$$= [\gamma^\mu\epsilon(1+\gamma^5)]_{ab}\partial_\mu = X_{ab}$$

say. Interchanging a and b gives

$$\{D_{La}, D_{Rb}\} = -((1+\gamma^5)\epsilon\gamma^{\mu T})_{ab}\partial_\mu$$

$$= -(\gamma^\mu\epsilon(1-\gamma^5))_{ab}\partial_\mu = X_{ba}$$

say. In matrix notation, these identities are expressible as

$$\{D_R, D_L^T\} = \gamma^\mu\epsilon(1+\gamma^5)\partial_\mu,$$

$$\{D_L, D_R^T\} = -\gamma^\mu\epsilon(1-\gamma^5)\partial_\mu$$

Adding these two equations and noting that D_L anticommutes with itself and D_R also anticommutes with itself, we get

$$\{D, D^T\} = 2\gamma^\mu\gamma^5\epsilon\partial_\mu$$

Then,

$$D_R^2 D_{La} = \epsilon_{bc}D_{Rb}D_{Rc}D_{La} =$$
$$\epsilon_{bc}D_{Rb}(X_{ca} - D_{La}D_{Rc})$$
$$= \epsilon_{bc}X_{ca}D_{Rb} - \epsilon_{bc}D_{Rb}D_{La}D_{Rc}$$
$$= \epsilon_{bc}X_{ca}D_{Rb} - \epsilon_{bc}(X_{ba} - D_{La}D_{Rb})D_{Rc}$$
$$= \epsilon_{bc}X_{ca}D_{Rb} - \epsilon_{bc}X_{ba}D_{Rc}$$
$$+D_{La}D_R^2$$

Equivalently,

$$[D_R^2, D_{La}] = \epsilon_{bc}X_{ca}D_{Rb} - \epsilon_{bc}X_{ba}D_{Rc}$$

Then,

$$D_R^2 D_{La}D_{Lp} = \epsilon_{bc}X_{ca}(X_{bp} - D_{Lp}D_{Rb})$$
$$-\epsilon_{bc}X_{ba}(X_{cp} - D_{Lp}D_{Rc}) + D_{La}([D_R^2, D_{Lp}] + D_{Lp}D_R^2)$$

so that

$$D_R^2 D_L^2 D_R^2 = \epsilon_{ap} D_R^2 D_{La} D_{Lp} D_R^2 =$$

$$= (epsilon_{bc} X_{ca} \epsilon_{ap} X_{bp} - \epsilon_{bc} X_{ba} \epsilon_{ap} X_{cp}) D_R^2$$

since the product of any three $D_R's$ is zero. We can express this relationship as

$$(D_R^2 D_L^2)^2 =$$

$$2Tr(\epsilon.X.\epsilon.X^T) D_R^2$$

and hence,

$$(D_L^2 D_R^2)^2 = 2Tr(\epsilon.X.\epsilon.X^T) D_L^2 D_R^2$$

Now,

$$X = \gamma^\mu \epsilon(1 + \gamma^5)\partial_\mu$$

and hence,

$$Tr(\epsilon.X.\epsilon.X^T) =$$

$$-Tr(\epsilon.\gamma^\mu \epsilon(1 + \gamma^5)^2 \epsilon^2 \gamma^{\nu T})\partial_\mu \partial_\nu$$

But,

$$-Tr(\epsilon.\gamma^\mu \epsilon(1 + \gamma^5)^2 \epsilon^2 \gamma^{\nu T})$$

$$2Tr(\epsilon.\gamma^\mu.\epsilon(1 + \gamma^5)\gamma^{\nu T})$$

$$= -2Tr(\epsilon \gamma^\mu \gamma^\nu (1 - \gamma^5)\epsilon)$$

$$= 2.Tr(\gamma^\mu \gamma^\nu (1 - \gamma^5)) = c.\eta^{\mu\nu}$$

where c is a real constant. Thus,

$$(D_L^2 D_R^2)^2 = c.\square.D_L^2 D_R^2$$

from which it follows that $P = c^{-1} D_L^2 D_R^2 / \square$ is idempotent, ie, a projection. Thus, the superpropagator satisfies

$$\square G = K\delta^4(\theta - \theta')\delta^4(x - x')$$

and hence,

$$G(x, \theta | x', \theta') = \Delta_F(x - x')\delta^4(\theta - \theta')$$

where $\Delta_F(x)$ is the Feynman propagator, ie,

$$\square \Delta_F(x) = \delta^4(x)$$

or equivalently, in terms of four dimensional Fourier transforms,

$$\mathcal{F}(\Delta_F)(k) = \int \Delta_F(x) exp(-ik.x) d^4x = \frac{1}{k^2 + i\epsilon},$$

where

$$k.x = k^0 x^0 - \sum_{r=1}^{3} k^r x^r, k^2 = k^{02} - \sum_{r=1}^{3} k^{r2}$$

[l] Design of quantum unitary gates using supersymmetric field theories:
Given a Lagrangian for a set of Chiral superfields and gauge superfields, we can construct the action as an integral of the Lagrangian over space-time. We can include forcing terms in this Lagrangian for example by adding c-number control gauge potentials to the quantum gauge field $V_\mu^A(x)$ or c-number control current terms to the terms involving the Dirac current which couples to the gauge field. After adding these c-number control terms, the resulting action will no longer be supersymmetric. However, we can still construct the Feynman path integral for the resulting action between two states of the field ie, by specifying the fields at the two endpoints of a time interval $[0,T]$ and then we obtain a transition matrix between these two states of the field. For example, the initial state can be a coherent state in which the annihilation component of the electromagnetic vector potential has definite values and the Dirac field of electrons and positrons is in a Fermionic coherent state where the annihilation component of the wave function has definite values. Likewise with the final state. Or else, we may specify the initial state to be a state in which there are definite numbers of photons, electrons and positrons having definite four momenta and spins and so also with the final state. In the case of a supersymmetric theory, we'll have to also specify the states of the other fields like the gaugino field, the gravitino field and the auxiliary fields or else we may break the supersymmetry by expressing the auxiliary fields in terms of the other superfield components using the variational equations of motion and then calculate the the Feynman path integral corresponding to an initial and a final state and then make this transition matrix as close as possible to a desired transition matrix by optimizing over the c-number control fields.

[m] Supersymmetry current.
Φ is an arbitrary left Chiral field. Consider the superfield

$$\Theta_\mu = \Phi^* \partial_\mu \Phi - \Phi \partial_\mu \Phi^* + a.\bar{D}_R \Phi^* \gamma_\mu D_L \Phi$$

where a is a suitably chosen constant. We recall that

$$\{D_R, D_L^T\} = \gamma^\mu \epsilon (1+\gamma^5) \partial_\mu$$

$$\{D_L, D_R^T\} = -\gamma^\mu \epsilon (1-\gamma^5) \partial_\mu$$

Also, with $D_R^2 = D_R^T \epsilon D_R$, we have seen that

$$[D_R^2, D_{La}] = \epsilon_{bc} X_{ca} D_{Rb} - \epsilon_{bc} X_{ba} D_{Rc}$$

where

$$X = \gamma^\mu \epsilon (1+\gamma^5) \partial_\mu$$

Thus,

$$[D_R^2, D_L] = -X^T \epsilon D_R - X^T \epsilon D_R = -2X^T \epsilon . D_R$$

$$= 2(1+\gamma^5)\gamma^\mu\epsilon\epsilon.\partial_\mu D_R = -2(1+\gamma^5)\gamma^\mu\partial_\mu D_R$$
$$= -2\gamma^\mu\partial_\mu(1-\gamma^5)D_R = -4\gamma^\mu\partial_\mu D_R$$
$$= -4(\gamma.\partial)D_R$$

Now recalling that

$$D^* = \gamma^5\epsilon\gamma^0 D$$

we had seen that

$$D_L^* = \gamma^5\epsilon\gamma^0 D_R, D_R^* = \gamma^5\epsilon\gamma^0 D_L$$

and hence, we get on conjugating the above commutation relation and using

$$(D_R^2)^* = D_R^{*T}\epsilon D_R^* = D_L^T\gamma^5\epsilon\gamma^0\epsilon\gamma^5\epsilon\gamma^0 D_L$$
$$= D_L^T\epsilon D_L = D_L^2$$

that

$$\gamma^5\epsilon\gamma^0[D_L^2, D_R] = -4\bar{\gamma}^\mu\partial_\mu\gamma^5\epsilon\gamma^0\partial_\mu D_L$$

or equivalently,

$$\epsilon\gamma^0[D_L^2, D_R] = 4\bar{\gamma}^\mu\epsilon\gamma^0\partial_\mu D_L$$

or noting that γ^0 and ϵ commute,

$$[D_L^2, D_R] = -4\epsilon\gamma^0\bar{\gamma}^\mu\epsilon\gamma^0\partial_\mu D_L$$

Now,

$$(\gamma^0\bar{\gamma}^\mu)^T = \gamma^{\mu*}\gamma^0 = (\gamma^0\gamma^\mu)^*$$
$$= \gamma^0\gamma^\mu$$

Thus,

$$\epsilon\gamma^0\bar{\gamma}^\mu = \epsilon\gamma^{\mu T}\gamma^0 = \gamma^\mu\epsilon\gamma^0$$

and hence, we finally get

$$[D_L^2, D_R] = 4\gamma^\mu\partial_\mu D_L = 4\gamma.\partial.D_L$$

Then,

$$(\gamma^\mu D_L)_a\Theta_\mu =$$
$$(-1)^{p(\Phi^*)}\Phi^*(\gamma^\mu\partial_\mu D_L)_a\Phi - \gamma^\mu D_L\Phi.\partial_\mu\Phi^* + a.(-1)^{p(\Phi^*)+1)}(\bar{D}_R\Phi^*(\gamma^\mu D_L)_a\gamma_\mu D_L\Phi)$$
$$+a((\gamma^\mu D_L)_a\bar{D}_R\Phi^*)\gamma_\mu.D_L\Phi$$

Now, the cth component of the term within the brackets in the last term of this equation is

$$(\gamma^\mu D_L)_a\bar{D}_{Rc}\Phi^* = \gamma^\mu_{ab}D_{Lb}\bar{D}_{Rc}\Phi^*$$
$$= \gamma^\mu_{ab}\{D_{Lb}, \bar{D}_{Rc}\}\Phi^*$$
$$= -\gamma^\mu_{ab}(\gamma^5\epsilon)_{cd}\{D_{Lb}, D_{Rd}\}\Phi^*$$
$$= \gamma^\mu_{ab}(\gamma^5\epsilon)_{cd}(\gamma^\nu\epsilon(1-\gamma^5))_{bd}\partial_\nu\Phi^*$$

$$= -(\gamma^\mu\gamma^\nu\epsilon(1-\gamma^5)\gamma^5\epsilon)_{ac}\partial_\nu\Phi^*$$
$$= -(\gamma^\mu\gamma^\nu(1-\gamma^5))_{ac}\partial_\nu\Phi^*$$

Thus, the last term evaluates to

$$a((\gamma^\mu D_L)_a\bar{D}_R\Phi^*)\gamma_\mu.D_L\Phi =$$

$$= -a\partial_\nu\Phi^*(\gamma^\mu\gamma^\nu(1-\gamma^5)\gamma_\mu D_L)_a\Phi$$
$$= -2a\partial_\nu\Phi^*(\gamma^\mu\gamma^\nu\gamma_\mu D_L)_a\Phi$$
$$= b.\partial_\nu\Phi^*(\gamma^\nu D_L)_a\Phi$$

Further,

$$D_{La}\bar{D}_R\Phi^*\gamma_\mu D_L\Phi = (-1)^{p(\Phi^*)+1)}\bar{D}_R\Phi^*\gamma^\mu D_{La}\gamma_\mu D_L\Phi$$

Thus,

$$(\gamma^\mu D_L)_a\bar{D}_R\Phi^*\gamma_\mu D_L\Phi =$$
$$(-1)^{p(\Phi^*)+1)}\bar{D}_R\Phi^*\gamma^\mu_{ab}\gamma_\mu D_{Lb}D_L\Phi$$

But,

$$\bar{D}_R\Phi^*\gamma^\mu_{ab}\gamma_\mu D_{Lb}D_L\Phi =$$
$$D_R^T\gamma^5\epsilon\Phi^*\gamma^\mu_{ab}\gamma_\mu D_{Lb}D_L\Phi =$$
$$= D_{Rc}\Phi^*(\gamma^5\epsilon\gamma_\mu)_{cd}\gamma^\mu_{ab}D_{Lb}D_{Ld}\Phi$$

Now, we can write

$$D_{Lb}D_{Ld} = D_L^T\epsilon D_L((1+\gamma^5)\epsilon/2)_{bd}$$

and hence we get

$$\bar{D}_R\Phi^*\gamma^\mu_{ab}\gamma_\mu D_{Lb}D_L\Phi =$$
$$D_{Rc}\Phi^*(\gamma^5\epsilon\gamma_\mu)_{cd}\gamma^\mu_{ab}(1+\gamma^5)\epsilon/2)_{bd}D_L^2\Phi$$
$$= -D_{Rc}\Phi^*(\gamma^\mu((1+\gamma^5)\epsilon/2)\epsilon\gamma_\mu\gamma^5)_{ac}D_L^2\Phi$$
$$= -(\gamma^\mu((1+\gamma^5)/2)\gamma_\mu)_{ac}D_{Rc}\Phi^*D_L^2\Phi$$
$$= -[\gamma^\mu(1+\gamma^5)/2)\gamma_\mu D_R]_a\Phi^*D_L^2\Phi$$
$$= -[\gamma^\mu\gamma_\mu D_R]_a\Phi^*D_L^2\Phi$$

Thus,

$$\gamma^\mu D_L\Theta_\mu =$$
$$(-1)^{p(\Phi^*)}\Phi^*\gamma^\mu\partial_\mu D_L\Phi - \gamma^\mu D_L\Phi.\partial_\mu\Phi^* + a.\gamma^\mu D_L(\bar{D}_R Phi^*\gamma_\mu D_L\Phi)$$
$$+b.\partial_\nu\Phi^*(\gamma^\nu D_L)_a\Phi$$
$$= (-1)^{p(\Phi^*)}\Phi^*\gamma^\mu\partial_\mu D_L\Phi - \gamma^\mu D_L\Phi.\partial_\mu\Phi^* + a.(-1)^{p(\Phi^*)+1)}(\bar{D}_R\Phi^*D_L^2\Phi)$$

$$+b.\partial_\nu\Phi^*(\gamma^\nu D_L)_a\Phi$$

$$= (-1/4)(-1)^{p(\Phi^*)}\Phi^* D_R D_L^2\Phi - \gamma^\mu D_L\Phi.\partial_\mu\Phi^* + c.(-1)^{p(\Phi^*)+1)} D_R\Phi^* D_L^2\Phi)$$

$$+b.\partial_\nu\Phi^*(\gamma^\nu D_L)_a\Phi$$

Now we recall the derivation of the superfield equations from the Lagrangian

$$[\Phi^*\Phi]_D + [f(\Phi)]_F = \int (D_R^2 S)^*(D_R^2 S)d^4x d^4\theta$$

$$+\int [f(D_R^2 S)]_F d^4x$$

We have with

$$f(D_R^2 S) = f(\Phi) = \sum_{r\geq 1} c(r)\Phi^r = \sum_{r\geq 1} c(r)(D_R^2 S)^r$$

that

$$\delta f(D_R^2 S) = \sum_r c(r)\delta((D_R^2 S)^r)$$

$$= \sum_r c(r)(D_R^2\delta S.(D_R^2 S)^{r-1} + D_R^2 S.D_R^2\delta S.(D_R^2 S)^{r-2} + ...)$$

$$= \sum_r c(r)D_R^2[(\delta S.(D_R^2 S)^{r-1}) + D_R^2 S.\delta S.(D_R^2 S)^{r-2} + ...]$$

$$= \sum_r c(r)D_R^2(\delta S.\delta f(\Phi)/\delta\Phi)$$

Thus,

$$\delta\int [f(D_R^2 S)]_F d^4x =$$

$$= \sum_r c(r)\int d^4x d^4\theta.[(\delta S.(D_R^2 S)^{r-1}) + D_R^2 S.\delta S.(D_R^2 S)^{r-2} + ...]$$

$$= \int \delta S.(\delta f(\Phi)/\delta\Phi)d^4x d^4\theta$$

Remark: Let S_e and S_o denote respectively the even and odd parts of any superfield S. Then,

$$D_R^2 S.\delta S.f(S) = (D_R^2 S_e + D_R^2 S_o)(\delta S_e + \delta S_o)f(S)$$

$$= [\delta S_e(D_R^2 S_e + D_R^2 S_o) + \delta S_o(D_R^2 S_e - D_R^2 S_o)]f(S)$$

Thus, we obtain the field equations

$$D_R^2\Phi^* = K.\delta f(\Phi)/\delta\Phi$$

which is equivalent to

$$D_L^2\Phi = K.(\delta f(\Phi)/\delta\Phi)^*$$

and hence,

$$\gamma^\mu D_L\Theta_\mu =$$

$$= (-1/4)(-1)^{p(\Phi^*)}\Phi^* D_R D_L^2\Phi - \gamma^\mu D_L\Phi.\partial_\mu\Phi^* + c.(-1)^{p(\Phi^*)+1)} D_R\Phi^* D_L^2\Phi)$$
$$+b.\partial_\nu\Phi^*(\gamma^\nu D_L)\Phi$$
$$= (-1)^{p(\Phi^*)+1}(b_1\Phi^* D_R(\delta f(\Phi/\delta\Phi)^* + b_2 D_R\Phi^*(\delta f(\Phi)/\delta\Phi)^*)$$
$$+b.\partial_\nu\Phi^*(\gamma^\nu D_L)\Phi - (-1)^{(p(\Phi)+1)p(\Phi^*)}\partial_\nu\Phi^*.\gamma^\nu D_L\Phi$$

Suppose instead we had defined Θ_μ as

$$\Theta_\mu = \Phi^*\partial_\mu\Phi - (-1)^{(p(\Phi)+1)p(\Phi^*)}\Phi\partial_\mu\Phi^* + a.\bar{D}_R\Phi^*\gamma_\mu D_L\Phi$$
$$= \Phi^*\partial_\mu\Phi - (-1)^{p(\Phi^*)}(\partial_\mu\Phi^*)\Phi + a.\bar{D}_R\Phi^*\gamma_\mu D_L\Phi$$

Then, we would have got

$$\gamma^\mu D_L\Theta_\mu =$$
$$= (-1)^{p(\Phi^*)+1}(b_1\Phi^* D_R(\delta f(\Phi/\delta\Phi)^* + b_2 D_R\Phi^*(\delta f(\Phi)/\delta\Phi)^*)$$
$$+b.\partial_\nu\Phi^*(\gamma^\nu D_L)\Phi - \partial_\nu\Phi^*.\gamma^\nu D_L\Phi$$

and by appropriate choice of a so that $b = 1$, the last two terms will cancel giving

$$\gamma^\mu D_L\Theta_\mu =$$
$$= (-1)^{p(\Phi^*)+1}(b_1\Phi^* D_R(\delta f(\Phi/\delta\Phi)^* + b_2 D_R\Phi^*(\delta f(\Phi)/\delta\Phi)^*)$$
$$= D_R\{(-1)^{p(\Phi^*)+1}c_1\Phi^*(\delta f(\Phi)/\delta\Phi)^* + c_2 f(\Phi)^*\}$$

Another interpretation of the supersymmetry current: Let $\alpha(x)$ be a space-time varying Majorana Fermionic parameter. Corresponding to this, we define an infinitesimal local supersymmetry transforation as $\bar{\alpha}(x)L$. Now, suppose $\mathcal{L}$ is a Lagrangian density function of component superfields such that its space-time integral yields a global supersymmetric action, ie it is invariant under $\bar{\alpha}L$ when α is a constant. Now, when $\alpha = \alpha(x)$ becomes a function of the space-time variables, the integral of $\mathcal{L}$ will no longer possess local symmetry. Its infinitesimal local supersymmetric variation will have generally two components, one, a component not containing space-time derivatives of $\alpha(x)$. This component can be written as $\int \bar{\alpha}(x)K^\mu_{,\mu}(x)d^4x$, two a component that contains only the first order partial derivatives of $\alpha(x)$ and is linear in these partial derivatives. Thus this component can be expressed as $\int \bar{\alpha}_{,\mu}(x)N^\mu(x)d^4x$. In case α is a constant parameter, both of these contributions will vanish. The vanishing of the first term can be attributed to global supersymmetry while the vanishing of the second can be attributed to Noether symmetry. In the general case of local supersymmetric transformations, the total change in the action is therefore the sum of these two components:

$$\delta S = \delta \int \mathcal{L} d^4x = \int (\bar{\alpha}(x) K^{\mu}_{,\mu}(x) + \alpha_{,\mu}(x) N^{\mu}(x)) d^4x$$

$$= \int \bar{\alpha}(x)(K^{\mu} - N^{\mu})_{,\mu}(x) d^4x$$

and in particular, when the field equations are satisfied, this variation must be zero for all $\alpha(x)$. This can happen only when

$$(K^{\mu} - N^{\mu})_{,\mu}(x) = 0$$

ie, when the supersymmetry current $S^{\mu} = K^{\mu} - N^{\mu}$ is conserved.

Calculation of the supercurrent for specific supersymmetric actions. In what follows we compute the supercurrent for three kinds of action with increasing degrees of complexity:

[1]

$$\mathcal{L} = [\Phi^* \Phi]_D + [f(\Phi)]_F$$

where Φ is an arbitrary left Chiral superfield. This Lagrangian describes only the dynamics of matter fields.

[2]

$$\mathcal{L} = [\Phi^* \Phi]_D + [\Phi^* V \Phi]_D + [f(\Phi)]_D + [W_L^T \epsilon W_L]_F$$

where V is a gauge superfield expressed in the Wess-Zumino gauge and

$$W_L = D_R^T \epsilon D_R D_L V$$

is the left Chiral Abelian gauge superfield. This Lagrangian describes the dynamics of super matter interacting with Abelian gauge and gaugino fields like the scalar and Dirac fields interacting with the electromagnetic field and the superpartner (Fermionic) of the photon.

[3]

$$\mathcal{L} = [\Phi^* exp(t.V)\Phi]_D + [f(\Phi)]_F + c_1 ReTr[W_L^T \epsilon W_L]_F + c_2 ImTr[W_L^T \epsilon W_L]_F$$

This Lagrangian describes the dynamics of supermatter interacting with non-Abelian gauge and gaugino fields.

[n] **Appendix Renormalization in quantum field theory**
Renormalization in quantum field theory.
[a] General graph theoretic considerations. Let I_f denote the number of internal lines of field type f. Let E_f denote the number of external field lines of

type f. Let N_i denote the number of vertices defined by interactions of type i. Let n_{if} denote the degree of the field of type f attached to a vertex of interaction type i. Note that each line carries a field of only one type by definition. Thus, n_{if} is the number of lines/edges of field type f that have one endpoint at a vertex of interaction type i. Then, it is clear that

$$E_f + 2I_f = \sum_i N_i n_{if}$$

The left side of this equation is the total number of times a vertex having a line of field type f attached to it occurs. If K lines of field type f are attached to a given vertex, then the contribution of that vertex to the lhs of this equation is K. Note that internal lines are attached to two vertices while external lines are attached to a single vertex. The rhs of the above equation also counts the number of vertices with multiplicities taken into account that a field line of type f is attached to it. n_{if} is the multiplicity of the vertex i corresponding to a field of type f, ie, the total number of lines of field type f that are attached to it.

[b] Let L denote the total number of loops in the connected graph. Then, Euler's theorem states that

$$V - E + L = 1$$

where

$$V = \sum_i N_i$$

is the total number of vertices, $E = \sum_f I_f$ is the number of edges, ie, internal lines.

[c] A propagator on an internal line of type f contributes a factor $D = 2s_f - 2$ to the power of the momentum (in the ultraviolet range) where $s_f = 0$ if the field type f is a photon and $s_f = 0$ if the field type f is an electron. This is because the photon propagator is $\eta_{mu\nu}/(k^2 + i\epsilon)$ while the electron propagator is $1/(\gamma.k - m + i\epsilon)$. If d_i field derivatives appear in a vertex of interaction type i, it then contributes a power $D = d_i$ to the power of the momentum. The integration measure d^4k associated with each loop of the graph contributes a momentum power of 4. Thus, the total power of momentum in the ultraviolet range of momentum of the graph is

$$D = 4L + \sum_f (2s_f - 2)I_f + \sum_i N_i d_i$$

$$= 4(1 + \sum_f I_f - \sum_i N_i) + \sum_f (2s_f - 2)I_f + \sum_i N_i d_i$$

$$= \sum_f (2s_f + 2)I_f + \sum_i N_i(d_i - 4) + 4$$

$$= \sum_f (s_f + 1)(\sum_i N_i n_{if} - E_f) + \sum_i N_i(d_i - 4) + 4$$

$$= 4 - \sum_f E_f(s_f + 1) - \sum_i N_i\Delta_i$$

where

$$\Delta_i = 4 - d_i - \sum_f n_{if}(s_f + 1)$$

If $\Delta_i \geq 0$ for all interactions i, then we get the bound

$$D \leq 4 - \sum_f E_f(s_f + 1)$$

no matter how many vertices of each interaction we may add. Hence, by redefining just a finite number of coupling constants, we can make the ultraviolet divergent integrals converge. However, if $\Delta_i < 0$ for some i, then by making the number of vertices of interaction type i, namely N_i larger and larger as it happens in the infinite Dyson series, D can be made indefinitely large and hence no finite number of redefinitions of the coupling constants can make this divergent integral converge. Thus, the condition for renormalizability of the field theory is that $\Delta_i \geq 0$ for all i.

[d] Examples. Let

$$\mathcal{L}_0 = (-1/4)F^{\mu\nu}F_{\mu\nu} + \bar{\psi}(i\gamma^\mu\partial_\mu - m)\psi$$

denote the sum of noninteracting photon and electron-positron field Lagrangians. The interaction Lagrangian is

$$\mathcal{L}_1 = eA_\mu\bar{\psi}.\gamma^\mu\psi$$

One of the terms in the perturbative computation of the corrected photon propagator using the Feynman path integral is

$$\int exp(i\int \mathcal{L}_0 d^4x)A_\alpha(x)A_\beta(y)(1/2!)(\int \bar{\psi}(x')\gamma^\mu\psi(x')\bar{\psi}(y')\gamma^\nu\psi(y')A_\mu(x')A_\nu(y')d^4x'd^4y')DA.D\psi.D\bar{\psi}$$

One of the terms in this computation is

$$\int D_{\alpha\mu}(x - x')D_{\beta\nu}(y - y')\gamma^\mu S_e(x' - y')\gamma^\nu S_e(y' - x')d^4x'd^4y'$$

where $D_{\mu\nu}$ is the photon propagator and S_e is the electron propagator. By using the Parseval relation, this double integral can be transformed into a double integral in the momentum domain and then the index μ defines one interaction vertex between the external photon line and two electron-positron lines while the index ν defines another interaction index, it is instructive to compute the behaviour of the integral for large values of the momentum. Specifically, the interaction term

$$\int \bar{\psi}(x)\gamma^\mu\psi(x)\bar{\psi}(y)\gamma^\nu\psi(y)A_\mu(x)A_\nu(y)d^4xd^4y$$

after transforming to the momentum domain there is a contribution of momentum degree -2 due to a photon propagator and a contribution of degree -2 due to two electron propagators.

As another example, let $\phi(x)$ be a real Klein-Gordon field and consider an interaction of this field with the photon field defined by the interaction action

$$\int \phi(x)(A^{\mu}(x)\partial_{\mu})^{p}\phi(x)d^{4}x$$

This can be expressed as

$$\int \phi(x)\partial_{\mu_1}...\partial_{\mu_p}\phi(x)A^{\mu_1}(x)...A^{\mu_p}(x)d^{4}x$$

While doing perturbative calculations, this interaction contributes a term

$$\int\int T\{\phi(u)\phi(v)\phi(x)\partial_{\mu_1}...\partial_{\mu_p}\phi(x)A^{\mu_1}(x)...A^{\mu_p}(x).\phi(y)\partial_{\mu_1}...\partial_{\mu_p}\phi(y)A^{\mu_1}(y)...A^{\mu_p}(y)\}d^{4}xd^{4}y$$

to the propagator correction for the Klein-Gordon field. Each derivative ∂_{μ_k} contributes a momentum factor p^{μ_k} in the momentum domain. So the contribution of this interaction vertex contributes a degree -6 for three Klein-Gordon field propagators and a degree $-4p$ for $2p$ photon propagators. After transforming to momentum space integrals using the Parseval theorem on Fourier transforms, the degree contributed by the momentum measures is $4(3+p-2) = 4(p+1)$. This is because two factors come by expressing the three Klein-Gordon propagators in the momentum domain using the Fourier transform, another p factors come while transforming the p photon propagator factors involving the electromagnetic four potential to the momentum domain and a factor of -2 comes because integration over $d^{4}xd^{4}y$ produces two momentum space delta functions which after integration over the momenta remove two four momenta integrals. Finally, the $2p$ space-time derivatives contribute a degree of $4(2p) = 8p$ of the momentum.

Remark: The contribution of this correction term to the KG propagator is given by

$$\int exp(i\int \mathcal{L}_0 d^{4}x)\phi(u)\phi(v)(\int \phi(x)(A^{\mu}(x)\partial_{\mu})^{p}\phi(x)d^{4}x)^{2}D\phi.DA$$

where

$$\mathcal{L}_0 = (-1/4)F^{\mu\nu}F_{\mu\nu} + (1/2)\partial^{\mu}\phi.\partial_{\mu}\phi$$

[e] If an integral of the form

$$I(q) = \int_{0}^{\infty} f(k,q)dk$$

diverges as $k \to \infty$, then we compute

$$d^n I(q)/dq^2 = \int_0^\infty \frac{\partial f(k,q)}{\partial q^n} dk$$

for the smallest value of n for which this converges. Denote it by $I_n(q)$. Then, $I(q)$ can be interpreted as

$$I(q) = \int^{(n)} I_n(q)dq + c_0 + c_1 q + ... + c_{n-1} q^{-n-1}$$

where $\int^{(n)} I_n(q)dq$ is an n-fold indefinite integral of $I_n(q)$ and $c_0, c_1, ..., c_{n-1}$ are infinite constants. This is one of the methods of renormalization.

[f] If an integral of the form $\int_\mu^\Lambda f(k,g)dk$ that represents a propagator correction or some scattering amplitude diverges as $\mu \downarrow 0$ (infrared divergence) of as $\Lambda \to \infty$ (ultraviolet divergence) or both where g is a coupling constant, or some parameter in the field theory Lagrangian like mass, charge or some parameter in the propagator, then we redefine a renormalized coupling constant $g_R = g_R(\mu, \Lambda)$ so that this integral assumes a definite finite value say c and then we express all the ampltiudes etc in terms of this renormalized parameter g_R:

$$c = \int_\mu^\Lambda f(k, g_R)dk$$

[g] Dimensional regularization. While calculating scattering amplitudes, propagator corrections etc., we may end up with divergent integrals of the form $\int f(k,p)d^4k$. In this case, the four momentum vector $k = (k^\mu : 0 \leq \mu \leq 3)$ is replaced by a d-momentum vector $k = (k^\mu : 0 \leq k \leq d-1)$ and when $d \neq 4$, this integral is found to converge to say $I(d)$. The singularity of $I(d)$ when $d = 4$ may appear in the form of a pole in d at $d = 4$ of order p, ie,

$$I(d) = I_0(p) + I_1/(d-4) + I_2/()d-4)^2 + ... + I_p/(d-4)^p$$

where $I_1, I_2, ..., I_p$ are constants. We then ignore these infinite constants $I_1, ..., I_p$ and retain only the term $I_0(p)$ in our result.

[h] Renormalization in qed.

[a] Vacuum polarization: The total Lagrangian of the electron-positron-photon field is given by

$$L = L_0 + L_1 + L_2 + L_3 + L_4 = (-1/4)F^B_{\mu\nu}F^{\mu\nu B} + \bar{\psi}^B(i\gamma^\mu\partial_\mu - m_B)\psi + e^B\bar{\psi}^B\gamma^\mu\psi^B A^B_\mu$$

where the superscript denotes the bare values of the fields, mass and charge B. The renormalized fields, mass and charge are defined without the superscript.

They are obtained by an appropriate scaling of the fields and charge and shift of the mass. These are

$$\psi = Z_2^{-1/2}\psi^B, A_\mu = Z_3^{-1/2}A_\mu^B, e = Z_3^{1/2}e^B, m = m_B + \delta m$$

Then, making these substitutions gives us

$$L = Z_3(-1/4)F_{\mu\nu}F^{\mu\nu} + Z_2\bar{\psi}(\gamma^\mu(i\partial_\mu + eA_\mu) - (m - \delta m))\psi$$

Thus, we may set

$$L_0 = (-1/4)F_{\mu\nu}F^{\mu\nu} + \bar{\psi}(i\gamma^\mu\partial_\mu - m)\psi,$$

$$L_1 = e\bar{\psi}\gamma^\mu\psi.A_\mu$$

$$L_2 = (Z_2 - 1)\bar{\psi}(i\gamma^\mu\partial_\mu - m)\psi + Z_2\delta m\bar{\psi}\psi,$$

$$L_3 = (Z_3 - 1)F_{\mu\nu}F^{\mu\nu}$$

$$L_4 = (Z_2 - 1)e\bar{\psi}\gamma^\mu\psi.A_\mu$$

In vacuum polarization, we evaluate the corrected photon propagator approximately, ie, upto one loop order:

$$\int exp(i\int Ld^4x)A_\mu(u)A_\nu(v)DA.D\psi.D\bar{\psi}$$

In a perturbative analysis of this treating L_1, L_2, L_3 as perturbations of L_0, one term that we encounter is

$$\int exp(i\int L_0d^4x)(-1/2)(\int L_1(z)d^4z)^2A_\mu(x)A_\nu(y)DA.D\psi.D\bar{\psi}$$

and this evaluates to in the momentum domain (ie after multiplication by $exp(iq.(u - v))$ followed by an integration over u to

$$D_{\mu\rho}(q)\Pi_1^{\rho\sigma}(q)D_{\sigma\nu}(q)$$

where $D_{\mu\nu}(q) = \eta_{\mu\nu}/(q^2 + i\epsilon)$ is the bare photon propagator, and

$$\Pi_1^{\rho\sigma}(q) = \int Tr(\gamma.p - m + i\epsilon)^{-1}\gamma^\rho.(\gamma.(p - q) - m + i\epsilon)^{-1}\gamma^\sigma)d^4p$$

After a Wick rotation followed by Feynman's trick, this evaluates to an expression of the form

$$\Pi_1^{\rho\sigma}(q) = f(d, m, q^2)(q^2\eta^{\rho\sigma} - q^\rho q^\sigma)$$

where dimension 4 has been replaced by d to make the integral converge. If $d = 4$, $f(d, m, q^2)$ becomes infinite. To get a finite result, we can introduce an infrared cutoff μ and also an ultraviolet cutoff Λ and then let $d \to 4$. This gives the f of the form a constant times $(\Lambda^{d-4} - \mu^{d-4})/(d - 4)$ which converges to $ln(\Lambda/\mu)$ as $d \to 4$. However, we can do without introducing these cutoffs by observing that the perturbation term L_3 to the Lagrangian gives another contribution to the photon propagator correction of the form

$$(Z_3 - 1)\int exp(i\int L_0 d^4x)A_\mu(x)A_\nu(y)(\int F_{\alpha\beta}(z)F^{\alpha\beta}(z)d^4z)DA.D\psi.D\bar{\psi}$$

which evaluates to an expression of the form

$$D_{\mu\rho}(q)\Pi_2^{\rho\sigma}(q)D_{\rho\nu}(q)$$

where

$$\Pi_2^{\rho\sigma}(q) = -(Z_3 - 1)(q^2\eta^{\rho\sigma} - q^\rho q^\sigma)$$

Thus, the total photon propagator correction upto one loop order is

$$D_{\mu\nu}\Pi^{\rho\sigma}D_{\sigma\nu}$$

where

$$\Pi^{\rho\sigma}(q) = \Pi_1^{\rho\sigma}(q) + \Pi_2^{\rho\sigma}(q) =$$

$$\pi(q^2)(q^2\eta^{\rho\sigma} - q^\rho q^\sigma)$$

with

$$\pi(q^2) = f(d, m, q^2) - (Z_3 - 1)$$

The infinity appearing in $f(d, m, q^2)$ when $d \to 4$ can be cancelled by choosing Z_3 to be an infinite constant appropriately once we note that the pole in $f(d, m, q^2)$ at $d = 4$ has a coefficient that is independent of q^2.

Complete photon propagator taking multiple loops into consideration: Inserting multiple loops of all order, the complete photon propagator is given by

$$D' = D + D\Pi.D + D\Pi.D\Pi.D + ... = D.(1 - D\Pi)^{-1}$$

$$= \frac{\eta}{q^2 + i\epsilon}.(1 - \frac{\eta.\pi(q^2)}{q^2 + i\epsilon})(\eta.q^2 - qq^T))^{-1}$$

$$= \frac{\eta}{q^2 + i\epsilon - \eta\pi(q^2)(\eta q^2 - qq^T)}$$

$$= \frac{\eta}{q^2(1 - \pi(q^2)) + \eta\pi(q^2)qq^T + i\epsilon}$$

The choice of Z_3 is made so that the pole as well as the residue at that pole remains the same as compared to that of the bare photon propagator (the pole of the bare photon propagator occurs when $q^2 = 0$). It is clear that these two conditions are satisfied iff $\pi(0) = 0$, or equivalently iff

$$Z_3 - 1 = f(d, m, 0)$$

[b] Schwinger's calculation of the anomalous magnetic moment of the electron. This deals with radiative correction to the electron's magnetic moment. To evaluate this, we must make the electron interact not only with the quantum photon field but also with an external classical photon field $\mathcal{A}_\mu$ and then evaluate the correction to the electron propagator produced not only by an internal quantum photon field that jumps over the electron line but also by a contribution from the external classical photon field which has a wavy line from the source to the middle of the electron line. The contribution of the external classical photon field to the total Lagrangian density of the electrons and photons is given by

$$L_5 = e\bar{\psi}\gamma^\mu\psi.\mathcal{A}_\mu$$

The vertex correction term to the electron propagator produced by this additional perturbation comes from the term

$$\int exp(i\int L_0 d^4x)(i\int L_5(z)d^4z).((-1/2)\int L_1(u)d^4u)^2\psi(x)\bar{\psi}(y)DAD\psi.D\bar{\psi}$$

There are totally eight Dirac field ψ terms in this path integral. Out of these eight terms, four of them will contribute end line electron propagators and the remaining four will contribute internal electron propagator. In this expression, there are also two quantum photon field A_μ terms which will pair up and contribute a photon propagator. The net result of this calculation will be a result of the form (in the momentum domain)

$$\int S_e(p_2,p_2')A_\mu(p_2'-p_1')\Gamma^\mu(p_2',p_1')S_e(p_1',p_1)d^4p_1'd^4p_2'$$

The vertex function is then defined by $\Gamma^\mu(p_2',p_1')$. We calculate the energy of the electron produced by this correction by replacing the end line electron propagators respectively by the incident and scattered electron wave functions u in the momentum domain. This energy contribution at momentum p is therefore given by

$$lim_{p'\to p}A_\mu(p'-p)\bar{u}(p')\Gamma^\mu(p',p)u(p)$$

There is yet another term that produces another correction to the electron propagator in addition to the vertex term. This involves introducing an electron loop in the middle of the quantum photon propagator appearing in the vertex diagram. This additional loop produces a vacuum polarization correction to the quantum photon propagator. Then, this total electron propagator correction $\mathcal{A}_\mu(p'-p)\Gamma^\mu(p',p)$ coming from the vertex terms having an external photon line has two components parametrized by two functions F and G of $(p'-p)^2$. In the limit as $p'\to p$ these functions converge to $F(0)$ and $G(0)$ respectively. Now, $G(q^2)$ is a finite integral while $F(q^2)$ is infinite. However, in order that the total charge density integrate out over the spatial volume to the electronic charge, we require that $F(0)+G(0)+1$. This is impossible since $F(0)$ is infinite. This problem is once again overcome by renormalization: We consider the contribution $(Z_2-1)e\bar{\psi}\gamma^\mu\psi.\mathcal{A}_\mu$ to the Lagrangian. This term contributes to the electron propagator a term

$$(Z_2-1)e\int \mathcal{A}_\mu(p_2'-p_1')S_e(p_2,p_2')\gamma^\mu S_e(p_1',p_1)d^4p_1'd^4p_2'$$

or equivalently, this represents a contribution of

$$(Z_2-1)e\gamma^\mu$$

to the vertex function. We can now choose Z_2 to be an infinite constant so that this term cancels out the infinity in the term $F(0)$ in Γ^μ. More precisely, $F(0)$ gets modified by this $e(Z_2-1)e\gamma^\mu$ term to $\tilde{F}(0)$ and we adjust the infinite constant Z_2 so that $\tilde{F}(0)+G(0)=1$. Then, $\tilde{F}(0)=1-G(0)$ determines the magnetic moment of the electron and hence $-G(0)$ determines Schwinger's correction to the magnetic moment of the electron.

[c] Electron self-energy.

Earlier, we had discussed the electron energy in the presence of a classical electromagnetic field using the vertex function and obtained thereby the anomalous magnetic moment as a contribution to the electron's energy caused by its interaction with both the quantum radiation field and the external electromagnetic field. Here to compute the electron's self energy, we ignore the external classical electromagnetic field and compute the one loop radiative corrections to the electron propagator which will yield its self energy. The main term in this electron propagation correction is

$$\int exp(i\int L_0d^4x)(-1/2)(\int L_1(u)d^4u)^2\psi(x)\bar{\psi}(y)DAD\psi.D\bar{\psi}$$

where

$$L_1=eA_\mu\bar{\psi}\gamma^\mu\psi$$

The Feynman diagram corresponding to this computation of the electron propagation correction is a straight line representing the electron propagator with a wavy photon line going from one end of the electron line to the other thereby forming and electron-photon loop. This diagram is the same as the diagram for the vertex function but with the external classical electromagnetic field line removed. We denote this correction in the momentum domain by

$$S_e(q)\Sigma_1(q)S_e(q)$$

which in the position domain reads

$$\int S_e(x,x')\Sigma_1(x',y')S_e(y',y)d^4xd^4y'$$

It is easy to verify from the diagram or by direct computation that

$$(\int L_1(u)d^4u)^2=e^2\int A_\mu(x')A_\nu(y')\bar{\psi}(x')\gamma^\mu\psi(x').\bar{\psi}(y')\gamma^\nu\psi(y')d^4x'd^4y'$$

and hence

$$\int S_e(x,x')\Sigma_1(x',y')S_e(y',y)d^4xd^4y' =$$

$$\int D_{\mu\nu}(x',y')S_e(x,x')\gamma^\mu S_e(x',y')\gamma^\nu S_e(y',y)d^4x'd^4y'$$

which yields

$$\Sigma_1(x',y') = D_{\mu\nu}(x',y')\gamma^\mu S_e(x',y').\gamma^\nu$$

or equivalently, by space-time homogeneity,

$$\Sigma_1(x'-y') = D_{\mu\nu}(x'-y')\gamma^\mu S_e(x'-y')\gamma^\nu$$

which gives on taking the four dimensional Fourier transform w.r.t $x'-y'$ a four-momentum space convolution term betweent the electron and photon propagators. The photon propagator is $O(k^{-2})$ while the electron propagator is $O(k^{-1})$ where k is the four momentum Hence, the convolution integral behaves as $O(k^{-1}) \times O(k^{-2})d^4k$ which is evidently ultraviolet divergent. To rectify it, we modify the photon propagator from $\frac{1}{k^2+i\epsilon}$ to

$$\frac{1}{k^2+i\epsilon} - \frac{1}{k^2-\mu^2+i\epsilon}$$

which behaves as $O(k^{-4})$ and its product with the electron propagator $O(k^{-1})$ followed by an integration w.r.t d^4k behaves as $\int k^{-5}.k^3dk = \int k^{-2}dk$ which is ultra violet convergent. This is the first step towards renormalization of the electron self energy. Finally, however we get an infinite result on letting $\mu \to \infty$. It should be noted that this computation yields $\Sigma_1(q)$ to be a function of $\gamma^\mu q_\mu$ only and also a function of the regulator mass μ. Now, we observe that there is a counter term in the form of the Lagrangian perturbation

$$L_2 = (Z_2-1)\bar{\psi}\gamma^\mu(i\partial_\mu - m)\psi + Z_2\delta m\bar{\psi}.\psi$$

which contributes an additional term

$$\int exp(i\int L_0 d^4x)i.(\int L_2(u)d^4u).\psi(x)\bar{\psi}(y)$$

which evaluates to

$$(Z_2-1)\int S_e(x-u)\gamma^\mu(i\partial_\mu - m)S_e(u-y)d^4u + Z_2\delta m\int S_e(x-u)S_e(u-y)d^4u$$

which in the momentum domain, ie, after four dimensional Fourier transforming w.r.t $x-y$ yields a contribution

$$S_(q)\Sigma_2(q)S_e(q)$$

where

$$\Sigma_2(q) = (Z_2-1)(\gamma^\mu q_\mu - m) + Z_2\delta m$$

Thus, the total corrected electron propagator upto one loop terms taking into account counter terms is given by

$$S_e(q)\Sigma(q)S_e(q)$$

where

$$\Sigma(q) = \Sigma_1(q, \mu) + \Sigma_2(q, Z_2)$$

where the explicit dependence of Σ_1 on the regularizing parameter μ and Σ_2 on Z_2 has been indicated. It should be noted that just as in the case of Σ_1, Σ_2 also depends on q only through the combination $\gamma^\mu q_\mu$. Just as in the case of the corrected photon propagator, the corrected electron propagator taking multiple loops into account is given by

$$S'_e(q) = S_e(q)(1 - \Sigma(q)S_e(q))^{-1}$$

where

$$S_e(q) = (\gamma^\mu q_\mu - m + i\epsilon)^{-1}$$

We shall choose Z_2 and δm so that the corrected electron propagator has a pole at the same point $\gamma^\mu q_\mu = m$ as the bare propagator and also has the same residue, namely unity. Now, writing $\gamma.q = \gamma^\mu q_\mu$, we have that

$$S'_e(q) = (\gamma.q - m + i\epsilon)^{-1}(1 - \Sigma(q)(\gamma.q - m + i\epsilon)^{-1})^{-1}$$

$$= (\gamma.q - m + i\epsilon - \Sigma(q))^{-1}$$

and this will have a pole at $\gamma.q = m$, provided that

$$\Sigma(q)|_{\gamma.q=m} = 0$$

Since

$$\Sigma_2(q)|_{\gamma.q=m} = Z_2\delta m$$

and hence we require that

$$\Sigma_1(q)|_{\gamma.q=m} = Z_2\delta m$$

for the location of the pole to remain unchanged and

$$\partial\Sigma_2(q)/\partial(\gamma.q)|_{\gamma.q=m} = 0$$

for the residue at that pole to remain unchanged. This second requirement yields

$$\partial\Sigma_1(q)/\partial(\gamma.q)|_{\gamma.q=m} = Z_2 - 1$$

After thus having obtained formulas for δm and Z_2 as functions of μ, we can easily show that the total electron propagation correction converges to a finite quantity as $\mu \to \infty$, ie,

$$\Sigma_1(q, \mu) + (Z_2(\mu) - 1)(\gamma^\mu q_\mu - m) + Z_2(\mu)\delta m(\mu)$$

converges to a finite quantity as $\mu \to \infty$. This defines the basic renormalization process for computing the electron self energy function.

Chapter 2

Some Aspects of Superstring Theory

[1] Discuss the irreducible representations of the permutation group using the group algebra method and Young diagrams. Explain how this theory can be used to obtain the characters of the permuatation group and how by using the duality between the action of the unitary group in its standard tensor representation and the corresponding action of the permutation group on tensors, one can derive a formula for the generating function for the characters of the permutation group provided that one uses Weyl's character formula for the characters of the unitary group.

[2] Superstring theory and supergravity coupled to the super Yang-Mills fields.

The super gravity field equations are determined by a Lagrangian density

$$L_{SUGR} = e[R^{mn}_{\mu\nu} e^{\mu}_{m} e^{\nu}_{n} + \bar{\chi}_a \Gamma^{abc} D_b \chi_c]$$

where Γ_{abc} is obtained by completely antisymmetrizing the product $\gamma_a\gamma_b\gamma_c$ of the Dirac matrices. χ_a are Majorana Fermion gravitino fields. Let ω^{mn}_{μ} denote the spin connection of the gravitational field. Then it defines a spinor covariant derivative by

$$D_\mu = \partial_\mu + \omega^{mn}_{\mu}\Gamma_{mn}, D_a = e^{\mu}_{a} D_\mu$$

where

$$\Gamma_{mn} = [\Gamma_m, \Gamma_n]$$

The Riemann curvature tensor in the spin representation is

$$R_{\mu\nu} = [\partial_\mu + \omega^{mn}_{\mu}\Gamma_{mn}, \partial_\nu + \omega^{rs}_{\nu}\Gamma_{rs}]$$

$$= [\omega^{mn}_{\nu,\mu} - \omega^{mn}_{\mu,\nu} + [\omega_\mu, \omega_\nu]^{mn}]\Gamma_{mn}$$

Note that $\{\Gamma_{mn}\}$ satisfy the Lorentz Lie algebra commutation relations and that we can write

$$R_{\mu\nu} = R^{mn}_{\mu\nu}\Gamma_{mn}$$

where

$$R^{mn}_{\mu\nu}e^{\alpha}_{m}e^{\beta}_{n} = R_{\mu\nu\alpha\beta}$$

is the standard Riemann-Christoffel curvature tensor in the coordinate basis. The curvature scalar is therefore

$$R = R^{mn}_{\mu\nu}e^{\mu}_{m}e^{\nu}_{n}$$

The supergravity Lagrangian is invariant, ie, changes by a perfect space-time divergence under the local supersymmetry transformation

$$\delta\chi_{\mu} = c_1 D_{\mu}\epsilon(x), \delta e^{a}_{\mu} = c_2\Gamma^{a}\chi_{\mu}(x)$$

where

$$\chi_a = e^{\mu}_{a}\chi_{\mu}$$

and

$$D_{\mu} = \partial_{\mu} + \omega^{ab}_{\mu}\Gamma_{ab}$$

is the gravitational spinor covariant derivative. The supersymmetry transformation of ω^{mn}_{μ} is not required here, since ω^{mn}_{μ} is assumed to be determined by the field equation that it satisfies obtained by setting the variational derivative of the supergravity action w.r.t. it to zero. This equation turns out to be a purely algebraic equation for ω^{mn}_{μ} which determines it in terms of the tetrad field e^{μ}_{n} and the gravitino field χ_a or equivalently χ_{μ}.

Remark: In the special relativistic Yang-mills theory, $F^{a}_{\mu\nu}$ are the fields obtained from the gauge Boson fields A^{a}_{μ} by

$$ieF^{a}_{\mu\nu} = [\partial_{\mu} + ieA_{\mu}, \partial_{\nu} + ieA_{\nu}]^{a}$$

with

$$A_{\mu} = A^{a}_{\mu}\tau_{a}$$

where the $\tau'_{a}s$ are Hermitian generators of the gauge group. A gauge invariant option for the Lagrangian density are obtained by adding matter action terms to the Yang-Mills Lagrangian is

$$L = (1/2)F^{a}_{\mu\nu}F^{\mu\nu a} + \bar{\psi}\Gamma^{a}[i\partial_{a} + eA_{a}) - m]\psi$$

$$+\bar{\eta}\Gamma^{ab}\eta B_{ab}$$

where

$$\Gamma^{ab} = [\Gamma^{a}, \Gamma^{b}]$$

It remains to determine the global supersymmetry transformations which change this Lagrangian by a total differential.

Exercise: Determine the above mentioned global supersymmetry transformation.

[3] Some additional details on supergravity:
Supergravity:
Let ω_μ^{mn} denote the spinor connection of the gravitational field. Then if Γ_m are the Dirac matrices in four dimensions and e_μ^m is the tetrad basis of space time being used, the covariant derivative of a spinor field is defined by

$$D_\mu \psi = (\partial_\mu + (1/4)\omega_\mu^{mn}\Gamma_{mn})\psi$$

where

$$\Gamma_{mn} = [\Gamma_m, \Gamma_n]$$

The curvature tensor in spinor notation is

$$R_{\mu\nu} = [\partial_\mu + (1/4)\omega_{\mu\nu}^{mn}\Gamma_{mn}, \partial_\nu + (1/4)\omega_\nu^{rs}\Gamma_{rs}]$$
$$= (1/4)(\omega_{\nu,\mu}^{mn} - \omega_{\nu,\mu}^{mn})\Gamma_{mn}$$
$$+(1/16)\omega_\mu^{mn}\omega_\nu^{rs}[\Gamma_{mn}, \Gamma_{rs}]$$

Now using the anticommutator

$$\{\Gamma_m, \Gamma_n\} = 2\eta_{mn}$$

we can easily show that

$$[\Gamma_{mn}, \Gamma_{rs}] = 4(\eta_{ms}\Gamma_{nr} + \eta_{nr}\Gamma_{ms} - \eta_{mr}\Gamma_{ns} - \eta_{ns}\Gamma_{mr})$$

Thus

$$R_{\mu\nu} =$$
$$= (1/4)(\omega_{\nu,\mu}^{mn} - \omega_{\nu,\mu}^{mn})\Gamma_{mn}$$
$$+(1/4)\omega_\mu^{mn}\omega_\nu^{rs}(\eta_{ms}\Gamma_{nr} + \eta_{nr}\Gamma_{ms} - \eta_{mr}\Gamma_{ns} - \eta_{ns}\Gamma_{mr})$$

This can be expressed as

$$R_{\mu\nu} = (1/4)R_{\mu\nu}^{mn}\Gamma_{mn}$$

where

$$R_{\mu\nu}^{mn} = \omega_{\nu,\mu}^{mn} - \omega_{\nu,\mu}^{mn} - \omega_\mu^{rn}\omega_\nu^{ms}\eta_{rs} + \omega_\mu^{ms}\omega_\nu^{rn}\eta_{sr} + \omega_\mu^{sn}\omega_\nu^{rm}\eta_{sr} - \omega_\mu^{mr}\omega_\nu^{ns}\eta_{rs}$$

=

$$\omega_{\nu,\mu}^{mn} - \omega_{\nu,\mu}^{mn} + \eta_{rs}(-\omega_\mu^{rn}\omega_\nu^{ms} + \omega_\mu^{ms}\omega_\nu^{rn} + \omega_\mu^{sn}\omega_\nu^{rm} - \omega_\mu^{mr}\omega_\nu^{ns})$$
$$= \omega_{\nu,\mu}^{mn} - \omega_{\nu,\mu}^{mn} + 2\eta_{rs}(\omega_\mu^{mr}\omega_\nu^{sn} - \omega_\nu^{mr}\omega_\mu^{ns})$$

It is easily shown that when the spinor connection ω_μ^{mn} for the gravitational field is appropriately chosen so that the Dirac equation in curved space-time remains invariant under both diffeormophisms and local Lorentz transformations, then the Riemann curvature tensor as defined usually in terms of the Christoffel connection symbols, coincides with $R_{\mu\nu\rho\sigma} = R^{mn}_{\mu\nu} e_{m\rho} e_{n\sigma}$. In particular, $R = R^{mn}_{\mu\nu} e^{\mu}_{m} e^{\nu}_{n}$ is the scalar curvature of space-time. The spinor connection ω^{mn}_{μ} is chosen so that the covariant derivative of the tetrad e^{n}_{μ} having one spinor index and one vector index is zero:

$$0 = D_{\nu} e^{n}_{\mu} = e^{n}_{\mu,\nu} - \Gamma^{\alpha}_{\mu\nu} e^{n}_{\alpha} + \omega^{nm}_{\nu} e_{m\mu}$$

This is an algebraic equation for ω^{mn}_{μ} and is easily solved. However when there are spinor fields like the gravitino in addition to the gravitational field specified by the tetrad e^{μ}_{n} (ie, the graviton), then the definition of the spinor connection has to be modified and it is expressed in terms of both the graviton and the gravitino fields. This equation is obtained by first considering the supergravity Lagrangian in four space-time dimensions

$$c_1 eR + i\bar{\chi}_{\mu} \Gamma^{\mu\nu\rho} D_{\nu} \chi_{\rho}$$

where χ_{μ} is a Majorana spinor having an additional vector index μ. The graviton tetrad field e^{μ}_{n} is Bosonic while the gravitino χ_{μ} is Fermionic. These are all considered in the quantum theory to be operator valued fields. Note the following self consistent definitions:

$$\Gamma^{\mu} = e^{\mu}_{n} \Gamma^{n}, \Gamma_{\mu} = g_{\mu\nu} \Gamma^{\nu} = \Gamma^{n} e_{n\mu}$$

$$\Gamma_{n} = \eta_{nm} \Gamma^{m}, e^{n}_{\mu} e_{n\nu} = g_{\mu\nu}, e^{\mu}_{n} e_{m\mu} = \eta_{nm},$$

$$e_{n\mu} = \eta_{nm} e^{m}_{\mu} = g_{\mu\nu} e^{\nu}_{n}$$

$$\Gamma^{\mu\nu\rho} = \Gamma^{\mu} \Gamma^{\nu\rho} + \Gamma^{\nu} \Gamma^{\rho\mu} + \Gamma^{\rho} \Gamma^{\mu\nu}$$

where

$$\Gamma^{\mu\nu} = [\Gamma^{\mu}, \Gamma^{\nu}]$$

Thus $\Gamma^{\mu\nu\rho}$ is obtained by antisymmetrizing the product $\Gamma^{\mu}\Gamma^{\nu}\Gamma^{\rho}$ over all the three indices. we can also clearly write

$$\Gamma^{\mu\nu} = e^{\mu}_{m} e^{\nu}_{n} \Gamma^{mn}, \Gamma^{\mu\nu\rho} = e^{\mu}_{m} e^{\nu}_{n} e^{\rho}_{k} \Gamma^{mnk}$$

In general, we can define

$$\Gamma^{\mu_1 \ldots \mu_k} = \sum_{\sigma \in S_n} sgn(\sigma) \Gamma^{\mu_{\sigma 1}} \ldots \Gamma^{\mu_{\sigma k}}$$

This is obtained by totally antisymmetrizing the product $\Gamma^{\mu_1} \ldots \Gamma^{\mu_k}$ over all its k indices. The basic property of a Majorana Fermionic operator field $\psi(x)$ is that apart from all its components anticommuting with each other, it has four components and satisfies

$$(\psi(x)^*)^T = \psi^T \Gamma^5 \epsilon \Gamma^0$$

where if

$$\psi(x) = \begin{pmatrix} \psi_1(x) \\ \psi_2(x) \\ \psi_3(x) \\ \psi_4(x) \end{pmatrix}$$

then

$$\psi(x)^* = \begin{pmatrix} \psi_1(x)^* \\ \psi_2(x)^* \\ \psi_3(x)^* \\ \psi_4(x)^* \end{pmatrix}$$

$\psi_k(x)^*$ denoting the operator adjoint of $\psi_k(x)$ in the Fock space on which it acts. Also we define

$$\psi(x)^T = [\psi_1(x), \psi_2(x), \psi_3(x), \psi_4(x)]$$

so that we have

$$(\psi(x)^*)^T = [\psi_1(x)^*, \psi_2(x)^*, \psi_3(x)^*, \psi_4(x)^*]$$

Now observe that

$$\epsilon = diag[i\sigma^2, i\sigma^2], \sigma^2 = \begin{pmatrix} 0 & -i \\ i & 0 \end{pmatrix}$$

Note that ϵ is a real skewsymmetric matrix. we write

$$e = i\sigma^2 = \begin{pmatrix} 0 & 1 \\ -1 & 0 \end{pmatrix}$$

so that

$$\epsilon = diag[e, e], e^2 = -I$$

$$\Gamma^5 \epsilon \Gamma^0 = \begin{pmatrix} e & 0 \\ 0 & -e \end{pmatrix} \begin{pmatrix} 0 & I \\ I & 0 \end{pmatrix}$$

$$= \begin{pmatrix} 0 & e \\ -e & 0 \end{pmatrix}$$

Thus, the condition for ψ to be a Majorana Fermion can be stated as

$$\psi_{1:2}^* = e\psi_{3:4}, \psi_{3:4}^* = -e\psi_{1:2}$$

or equivalently,

$$(\psi_{1:2}^*)^T = -(\psi_{3:4}^*)^T e, (\psi_{3:4}^*)^T = (\psi_{1:2}^*)^T e$$

Also,

$$\Gamma^0\Gamma^n, \Gamma^0\Gamma^\mu, \Gamma^n\Gamma^0, \Gamma^\mu\Gamma^0$$

are Hermitian matrices. We observe that if ψ is a Majorana Fermion,

$$\bar{\psi} = (\psi^*)^T\Gamma^0 = \psi^T\Gamma^5\epsilon\Gamma^0$$

So we can also write down the Lagrangian of the gravitino as

$$i\bar{\chi}_\mu\Gamma^{\mu\nu\rho}D_\nu\chi_\rho$$

$$= i\chi_\mu^{*T}\Gamma^0\Gamma^{\mu\nu\rho}D_\nu\chi_\rho$$

$$= i\chi_\mu^T\Gamma^5\epsilon\Gamma^0\Gamma^{\mu\nu\rho}D_\nu\chi_\rho$$

We can verify that apart from a perfect divergence, this quantity is a Hermitian operator. First observe that

$$(\Gamma^0\Gamma^\mu\Gamma^\nu\Gamma^\rho)^* =$$

$$(\Gamma^0\Gamma^\mu\Gamma^\nu\Gamma^0\Gamma^0\Gamma^\rho)^* =$$

$$\Gamma^0\Gamma^\rho\Gamma^\nu\Gamma^0\Gamma^0\Gamma^\mu$$

$$= \Gamma^0\Gamma^\rho\Gamma^\nu\Gamma^\mu$$

so that on antisymmetrizing over the three indices, we get

$$(\Gamma^0\Gamma^{\mu\nu\rho})^* = -\Gamma^0\Gamma^{\mu\nu\rho}$$

Thus,

$$(i\chi_\mu^{*^T}\Gamma^0\Gamma^{\mu\nu\rho}\partial_\nu\chi_\rho)^*$$

$$= i\chi_{\rho,\nu}^{*T}\Gamma^0\Gamma^{\mu\nu\rho}\chi_\mu$$

$$= i\chi_{\mu,\nu}^{*T}\Gamma^0\Gamma^{\rho\nu\mu}\chi_\rho$$

$$= -i\chi_{\mu,\nu}^{*T}\Gamma^0\Gamma^{\mu\nu\rho}\chi_\rho$$

$$= \partial_\nu(-i\chi_\mu^{*T}\Gamma^0\Gamma^{\mu\nu\rho}\chi_\rho)$$

$$+ i\chi_\mu^{*T}\Gamma^0\Gamma^{\mu\nu\rho}\chi_{\rho,\nu}$$

proving our claim provided that we replace D_ν by ∂_ν. If we take the connection into account, ie

$$D_\nu\chi_\rho = \partial_\nu\chi_\rho + (1/4)\omega_\nu^{mn}\Gamma_{mn}\chi_\rho$$

$$-\Gamma^\alpha_{\rho\nu}\chi_\alpha$$

then it follows that we must prove the Hermitianity of the operator fields

$$\chi_\mu^{*T}\Gamma^0\Gamma^{\mu\nu\rho}\Gamma_{mn}\chi_\rho.\omega_\nu^{mn} --- (a)$$

and

$$\chi_\mu^{*T}\Gamma^0\Gamma^{\mu\nu\rho}\chi_\alpha.\Gamma^\alpha_{\rho\nu} --- (b)$$

However the field (b) is identically zero since $\Gamma^{\mu\nu\rho}$ is antisymmetric in (ν,ρ) while $\Gamma^{\alpha}_{\rho\nu}$ is symmetric in (ν,ρ). Hence, we have to prove only the Hermitianity of the field

$$\chi_{\mu}^{*T}\Gamma^{0}\Gamma^{\mu\nu\rho}\Gamma_{mn}\chi_{\rho}---(c)$$

Now,

$$\Gamma^{\mu\nu\rho}\Gamma_{mn}=[\Gamma^{\mu\nu\rho},\Gamma_{mn}]+\Gamma_{mn}\Gamma^{\mu\nu\rho}$$

and

$$[\Gamma_{pqr},\Gamma_{mn}]=[\Gamma_{p}\Gamma_{qr}+\Gamma_{q}\Gamma_{rp}+\Gamma_{r}\Gamma_{pq},\Gamma_{mn}]$$

Now,

$$[\Gamma_{p}\Gamma_{qr},\Gamma_{mn}]=\Gamma_{p}[\Gamma_{qr},\Gamma_{mn}]+[\Gamma_{p},\Gamma_{mn}]\Gamma_{qr}$$

$$=4\Gamma_{p}(\eta_{qn}\Gamma_{rm}+\eta_{rm}\Gamma_{qn}-\eta_{qm}\Gamma_{rn}-\eta_{rn}\Gamma_{qm})$$

$$+4(\eta_{pm}\Gamma_{n}-\eta_{pn}\Gamma_{m})\Gamma_{qr}$$

Summing this equation over cyclic permutations of (pqr) gives us

$$[\Gamma_{pqr},\Gamma_{mn}]=$$

$$4\sum_{(pqr)}\eta_{mq}(\Gamma_{p}\Gamma_{nr}+\Gamma_{r}\Gamma_{pn}+\Gamma_{n}\Gamma_{rp})$$

$$+4\sum_{(pqr)}\eta_{nq}(\Gamma_{p}\Gamma_{rm}+\Gamma_{r}\Gamma_{mp}+\Gamma_{m}\Gamma_{pr})$$

$$=4\sum_{(pqr)}(\eta_{mq}\Gamma_{pnr}+\eta_{nq}\Gamma_{prm})$$

Note that this quantity is antisymmetric w.r.t interchange of (m,n). It thus follows that

$$e_{q\nu}\chi_{\mu}^{*T}\Gamma^{0}[\Gamma^{\mu\nu\rho},\Gamma_{mn}]\chi_{\rho}$$

$$=\chi^{p*T}\Gamma^{0}[\Gamma_{pqr},\Gamma_{mn}]\chi^{r}$$

$$=4\sum_{(pqr)}[\eta_{mq}\chi^{p*T}\Gamma^{0}\Gamma_{pnr}\chi^{r}+\eta_{nq}\chi^{p*T}\Gamma^{0}\Gamma_{prm}\chi^{r}]$$

Now,

$$(\chi^{p*T}\Gamma^{0}\Gamma_{pnr}\chi^{r})^{*}=$$

$$\chi^{r*T}\Gamma^{0}\Gamma^{rnp}\chi^{p}$$

$$=\chi^{p*T}\Gamma^{0}\Gamma^{pnr}\chi^{r}$$

which proves the Hermitianity. Another way to see the Hermitianity of this is to use the Majorana Fermion property of χ^{p} to write

$$\chi^{p*T}\Gamma^{0}\Gamma_{pnr}\chi^{r}=$$

$$\chi^{pT}\Gamma^{5}\epsilon\Gamma_{pnr}\chi^{r}$$

and use the fact that

$$(\Gamma^5\epsilon\Gamma_{pnr})^T = -\Gamma_{pnr}^T\epsilon\Gamma^5$$

$$= -\epsilon\Gamma_{rnp}\Gamma^5 = \Gamma^5\epsilon\Gamma_{rnp} = -\Gamma^5\epsilon\Gamma_{pnr}$$

where we have used the identities

$$\Gamma_n^T\epsilon = \epsilon\Gamma_n, \Gamma_n\Gamma^5 = -\Gamma^5\Gamma_n$$

Now consider

$$X = \chi_\mu^{*T}\Gamma^0\{\Gamma^{\mu\nu\rho}, \Gamma_{mn}\}\chi_\rho$$

where $\{.,.\}$ denotes anticommutator. We have

$$X = X_1 + X_2$$

where

$$X_1 = \chi_\mu^{*T}\Gamma^0\Gamma^{\mu\nu\rho}\Gamma_{mn}\chi_\rho$$

$$X_2 = \chi_\mu^{*T}\Gamma^0\Gamma_{mn}\Gamma^{\mu\nu\rho}\chi_\rho$$

we have

$$X_1^* = \chi_\rho^{*T}\Gamma^0\Gamma_{mn}\Gamma^{\mu\nu\rho}\chi_\mu$$

$$= -\chi_\mu^{*T}\Gamma^0\Gamma_{mn}\Gamma^{\mu\nu\rho}\chi_\rho = -X_2$$

which shows that

$$X^* = -X$$

ie, X is skew Hermitian. Note that we have used the fact that

$$(\Gamma^0\Gamma_p\Gamma_q\Gamma_r\Gamma_m\Gamma_n)^* =$$

$$(\Gamma^0\Gamma_p\Gamma^0\Gamma^0\Gamma_q\Gamma_r\Gamma^0\Gamma^0\Gamma_m\Gamma^0\Gamma^0\Gamma_n)^* =$$

$$\Gamma^0\Gamma_n\Gamma_m\Gamma^0\Gamma^0\Gamma_r\Gamma^0\Gamma^0\Gamma_q\Gamma_p\Gamma^0\Gamma^0 =$$

$$\Gamma^0\Gamma_n\Gamma_m\Gamma_r\Gamma_q\Gamma_p$$

since $\Gamma^0\Gamma_n, \Gamma_n\Gamma^0, \Gamma^0$ are Hermitian and $\Gamma^{02} = I$. Thus, by antisymmetrizing over (pqr) and over (mn), we get

$$(\Gamma^0\Gamma_{pqr}\Gamma_{mn})^* = \Gamma^0\Gamma_{nm}\Gamma_{rqp}$$

$$= \Gamma^0\Gamma_{mn}\Gamma_{pqr}$$

Now consider the following local supersymmetry transformation

$$\delta\chi_\mu(x) = D_\mu\epsilon(x), \delta e_\mu^n = K\bar{\epsilon}(x)\Gamma^n\chi_\mu(x)$$

where $\epsilon(x)$ is an infinitesimal Majorana Fermionic parameter. We can easily check that $D_\mu\epsilon$ also satisfies the Majorana Fermion property. Indeed,

$$((D_\mu\epsilon(x))^*)^T = \partial_\mu(\epsilon(x)^*)^T + (\epsilon(x)^*)^T\Gamma^*_{mn}\omega_\mu^{mn}$$

$$= (\partial_\mu\epsilon(x))^T\Gamma^5\epsilon\Gamma^0 + \epsilon(x)^T\Gamma^5\epsilon\Gamma^0\Gamma^*_{mn}\omega_\mu^{mn}$$

$$= (\partial_\mu\epsilon(x))^T\Gamma^5\epsilon\Gamma^0 - \epsilon(x)^T\Gamma^5\epsilon\Gamma_{mn}\Gamma^0\omega_\mu^{mn}$$

Now,

$$\epsilon\Gamma_{mn} = -\Gamma_{mn}^T\epsilon$$

since

$$\Gamma_n^T\epsilon = \epsilon\Gamma_n$$

Thus, since $\Gamma^{0T} = \Gamma^0$, it follows that

$$\Gamma^5\epsilon\Gamma_{mn}\Gamma^0 = \Gamma^5\epsilon\Gamma_{mn}\Gamma^0 =$$

$$-\Gamma^5\Gamma_{mn}^T\epsilon\Gamma^0 = -\Gamma_{mn}^T\Gamma^5\epsilon\Gamma^0$$

his gives

$$((D_\mu\epsilon(x))^*)^T =$$

$$(\partial_\mu\epsilon(x))^T\Gamma^5\epsilon\Gamma^0$$

$$+\epsilon(x)^T\Gamma_{mn}^T\Gamma^5\epsilon\Gamma^0$$

$$= (D_\mu\epsilon(x))^T\Gamma^5\epsilon\Gamma^0$$

proving thereby the Majorana property of $D_\mu\epsilon(x)$. Now under the local supersymmetry transformation of χ_μ, the Gravitino Lagrangian changes by

$$\delta_\chi(\bar{\chi}_\mu\Gamma^{\mu\nu\rho}D_\nu\chi_\rho) =$$

$$= \delta\bar{\chi}_\mu\Gamma^{\mu\nu\rho}D_\nu\chi_\rho +$$

$$\bar{\chi}_\mu\Gamma^{\mu\nu\rho}D_\nu\delta\chi_\rho$$

$$= \bar{D}_\mu\epsilon(x)\Gamma^{\mu\nu\rho}D_\nu\chi_\rho$$

$$+\bar{\chi}_\mu\Gamma^{\mu\nu\rho}D_\nu D_\rho\epsilon(x)$$

The term in this quantity that is quadratic in $\{\omega_\mu^{mn}\}$ is given by

[4] Dynamics of superparticles.
The action integral of a superparticle is given by

$$S = \int e.p_\mu p^\mu d\tau$$

where e is the square root of a one dimensional metric and

$$p^\mu = dx^\mu/d\tau - \theta^A\epsilon\gamma^\mu d\theta^A/d\tau$$

with $\theta^A(\tau)$ being Fermionic coordinates. The sum is over A and each θ^A is therefore a D dimensional Majorana Fermion. We wish to determine the supersymmetry transformation under which S is invariant. We consider an infinitesimal supersymmetry transformation

$$\delta\theta^A = \gamma^0\gamma^\mu p_\mu.k^A = \alpha^\mu p_\mu.k^A$$

$$\delta x^\mu = \theta^{AT}\epsilon\gamma^\mu\delta\theta^A$$

where k^A are infinitesimal Grassmanian vectors, each of size $D \times 1$. Note that $\epsilon\gamma^\mu$ is a skewsymmetric matrix. Then,

$$\delta p^\mu = (\delta x^\mu)' - \delta\theta^{AT}\epsilon\gamma^\mu\theta^{A'}$$

$$-\theta^{AT}\epsilon\gamma^\mu\delta\theta^{A'}$$

$$= \theta^{AT'}\epsilon\gamma^\mu\delta\theta^A$$

$$\theta^{AT}\epsilon\gamma^\mu\delta\theta^{A'}$$

$$-\delta\theta^{AT}\epsilon\gamma^\mu\theta^{A'}$$

$$-\theta^{AT}\epsilon\gamma^\mu\delta\theta^{A'}$$

$$= \theta^{AT'}\epsilon\gamma^\mu\delta\theta^A - \delta\theta^{AT}\epsilon\gamma^\mu\theta^{A'}$$

$$= 2\theta^{AT'}\epsilon\gamma^\mu\delta\theta^A$$

since k^B and hence $\delta\theta^B$ commutes with θ^A and $\epsilon\gamma^\mu$ is skewsymmetric. We thus get

$$\delta(p^2) = 2p_\mu\delta p^\mu =$$

$$4\theta^{AT'}\epsilon\gamma^\mu p^\mu\delta\theta^A$$

$$= 4\theta^{AT'}\epsilon\gamma^0(\alpha.p)\delta\theta^A$$

$$= 4\theta^{AT'}\epsilon\gamma^0(\alpha.p)^2 k^A$$

$$= 4p^2\theta^{AT'}\epsilon\gamma^0 k^A$$

and hence

$$\delta(e.p^2) = \delta(e)p^2 + e\delta(p^2)$$

$$= \delta(e)p^2 + 4ep^2\theta^{AT'}\epsilon\gamma^0 k^A$$

This is zero provided that under the supersymmetry transformation, e changes by

$$\delta e = -4e.\theta^{AT'}\epsilon\gamma^0 k^A$$

We note that in this analysis, we are not assuming that k^A is independent of time τ. It can depend on τ and hence our action is not only globally supersymmetric but it is also locally supersymmetric.

[5] Generalization of supersymmetric actions from super-particles to super-strings. We define

$$p^{\mu}_{\alpha} = \partial_{\alpha}X^{\mu} - \theta^{AT}\epsilon\gamma^{\mu}\partial_{\alpha}\theta^{A}$$

where $\alpha = 1, 2$ and

$$\partial_1 = \partial_{\tau}, \partial_2 = \partial_{\sigma}$$

We define the infinitesimal super-symmetric transformations

$$\delta\theta^{A} = k^{A}, \delta X^{\mu} = k^{AT}\epsilon\gamma^{\mu}\theta^{A}$$

where k^{A} is an infinitesimal Grassmanian parameter. Then,

$$\delta p^{\mu}_{\alpha} = (\delta X^{\mu})_{,\alpha} - \delta\theta^{AT}\epsilon\gamma^{\mu}\partial_{\alpha}\theta^{A} - \theta^{AT}\epsilon\gamma^{\mu}(\delta\theta^{A})_{,\alpha}$$

$$= k^{AT}_{,\alpha}\epsilon\gamma^{\mu}\theta^{A} + k^{AT}\epsilon\gamma^{\mu}\theta^{A}_{,\alpha} - k^{AT}\epsilon\gamma^{\mu}\theta^{A}_{,\alpha} - \theta^{AT}\epsilon\gamma^{\mu}k^{A}_{,\alpha}$$

$$= 2k^{AT}_{,\alpha}\epsilon\gamma^{\mu}\theta^{A}$$

This is zero iff k^{A} is a constant Grassmanian parameter in which case it follows that the action

$$S = \int \eta_{\mu\nu}h^{\alpha\beta}\sqrt{h}p^{\mu}_{\alpha}p^{\nu}_{\beta}d^{2}\sigma$$

is supersymmetric provided that the world sheet metric $h^{\alpha\beta}$ is assumed to be supersymmetric invariant. Thus the action has only global supersymmetry and not local supersymmetry. To obtain local supersymmetry, we have to add other terms to the action. (See M.Green,J.Schwarz and E.Witten, Superstring Theory, vol.I, Cambridge University Press).

[6] Super Yang-Mills action.

$$L = K_1 F^{a}_{\mu\nu}F^{\mu\nu a} + K_2\psi^{aT}\epsilon\gamma^{\mu}D_{\mu}\psi^{a}$$

where

$$iF_{\mu\nu} = [\partial_{\mu} + iA_{\mu}, \partial_{\nu} + iA_{\nu}]$$

or equivalently,

$$F^{a}_{\mu\nu} = A^{a}_{\nu,\mu} - A^{a}_{\mu,\nu} + C(abc)A^{b}_{\mu}A^{c}_{\nu}$$

where $C(abc)$ are the structure constants of a set of Hermitian basis for the Lie algebra of the gauge group. D_{μ}, the gauge covariant derivative acts in the adjoint representation on the gaugino fields ψ^{a}:

$$D_{\mu}\psi^{a} = \partial_{\mu}\psi^{a} + iC(abc)A^{b}_{\mu}\psi^{c}$$

This formula can be obtained using

$$D_{\mu}\psi^{a} = [\partial_{\mu} + iA_{\mu}, \psi]^{a} =$$

$$\partial_{\mu}\psi^{a} + i[A^{b}_{\mu}\tau_{b}, \psi^{c}\tau_{c}]^{a}$$

$$= \partial_\mu \psi^a + C(abc) A_\mu^b \psi^c$$

Note that A_μ^a is a gauge Boson field and its superpartner ψ^a is a gauge Fermion field, also called a gaugino field. Now, we must introduce infinitesimal supersymmetry transformations under which the action $\int L d^4 x$ is invariant. We assume that such a transformation has the form

$$\delta A_\mu^a = k^T \epsilon \gamma_\mu \psi^a,$$

$$\delta \psi^a = (\gamma^{\mu\nu} k) F_{\mu\nu}^a$$

where k is an infinitesimal Grassmannian four component spinor and $\gamma^{\mu\nu} = [\gamma^\mu, \gamma^\nu]$. Under this transformation, we get

$$\delta(F_{\mu\nu}^a F^{\mu\nu a}) = 2F^{\mu\nu a} \delta F_{\mu\nu}^a =$$

$$= 2F^{\mu\nu a}(\delta A_{\nu,\mu}^a - \delta A_{\mu,\nu}^a + C(abc)(A_\mu^b \delta A_\nu^c + A_\nu^c \delta A_\mu^b))$$

$$= 4F^{\mu\nu a}(\delta A_{\nu,\mu}^a + C(abc) A_\mu^b \delta A_\nu^c)$$

$$\delta(\psi^{aT} \epsilon \gamma^\mu D_\mu \psi^a) =$$

$$\delta \psi^{aT} \epsilon \gamma^\mu D_\mu \psi^a$$

$$+ \psi^{aT} \epsilon \gamma^\mu D_\mu \delta \psi^a$$

$$+ \psi^{aT} \epsilon \gamma^\mu (\delta D_\mu) \psi^a$$

where

$$(\delta D_\mu) \psi^a = C(abc)(\delta A_\mu^b) \psi^c$$

Exercise: Determine the constraint on the Dirac Gamma matrices that ensure local supersymmetry of the above super-Yang-Mills action. Show that local supersymmetry arises at certain critical dimensions of the Dirac Gamma matrices.

[7] **Dirac's equation in curved space-time**

Let $\mathcal{M}$ be a Riemannian manifold with metric g, ie, for each pair of vector fields X, Y on $\mathcal{M}$, we have a scalar field $g(X, Y)$, such that the map $(X, Y) \rightarrow g(X, Y)$ is symmetric and bilinear. We are given a connection ∇ on $\mathcal{M}$. We say that ∇ is induced by the metric if $\nabla_X g = 0$ for every vector field X defined on $\mathcal{M}$. This is equivalent to saying that

$$X(g(Y, Z)) = g(\nabla_X Y, Z) + g(Y, \nabla_X Z)$$

for any three vector fields X, Y, Z. It is not hard to prove that the connection ∇ induced by the metric is uniquely determined by the metric provided we assume in addition that the torsion of the connection is zero, ie,

$$\nabla_X Y - \nabla_Y X - [X, Y] = 0$$

for any two vector fields X, Y. An elliptic differential operator on the Riemannian manifold is a second order differential operator of the form

$$D = g^{ij}(x)\partial_i\partial_j + c^i(x)\partial_i + b(x)$$

where we have chosen and fixed the coordinate system x and have defined $g^{ij}(x)$ so that

$$g^{ij}(x) = ((g_{ij}(x)))^{-1}$$

where

$$g_{ij}(x)X^i(x)Y^j(x) = g(X,Y)(x)$$

A Dirac operator on the Riemannian manifold is an operator of the form

$$D = V_a^\mu(x)\gamma^a(\partial_\mu + ieA_\mu^b(x)\tau_b + \Gamma_\mu(x)) - m$$

where $V_a^\mu(x)$ is the Vierbein of the metric $g_{\mu\nu}(x)$. We can define the space-time dependent Dirac matrices

$$\Gamma^\mu(x) = V_a^\mu(x)\gamma^a$$

Then, we can define the Dirac operator as

$$D = \Gamma^\mu(x)(i\partial_\mu + eA_\mu(x)) - m$$

where now $A_\mu(x)$ is a matrix valued four vector potential. It takes into account both the Yang-Mills connection $A_\mu^b(x)\tau_b$ and the spinor connection

$$\Gamma_\mu(x) = \omega_{ab}^\mu(x)[\gamma^a, \gamma^b]$$

The Dirac operator D acts on the space of N-component spinor fields defined on the manifold $\mathcal{M}$.

[8] **Integration on a differentiable manifold**

Let $\mathcal{M}$ be an n-dimensional manifold and ω a $k \le n$ differential form on $\mathcal{M}$. Stokes' theorem states that

$$\int_{\partial M} \omega = \int_{\mathcal{M}} d\omega$$

In terms of coordinates,

$$\omega(x) = \omega_{\mu_1 \ldots \mu_k}(x)dx^{\mu_1} \wedge \ldots \wedge dx^{\mu_k}$$

Then,

$$d\omega(x) = \omega_{\mu_1 \ldots \mu_k, m}(x)dx^m \wedge dx^{\mu_1} \wedge \ldots \wedge dx^{\mu_k}$$

Then

$$\int_{\mathcal{M}} d\omega = \sum_{m,\mu_1,\ldots,\mu_k distinct and in increasing order} sgn(m, mu_1, \ldots, \mu_k)$$

$$\int \omega_{\mu_1 \ldots \mu_k, m}(x)dx^m dx^{\mu_1} \ldots dx^{\mu_k}$$

where $sgn(m, \mu_1, ..., \mu_m)$ is the signature of the permutation that takes the sequence $(m, \mu_1, ..., \mu_k)$ to the sequence obtained by arranging $m, \mu_1, ..., \mu_k$ in increasing order.

[9] **Index of an Elliptic operator**

Now, let P be an elliptic differential operator. Consider the heat kernel

$$K_t = exp(-tP)$$

Let $K_t(x, y)$ denotes its kernel, ie,

$$\int K_t(x, y)f(y)dy = (exp(-tP))f(x)$$

We have

$$Tr(exp(-tP)) = \int_{\mathcal{M}} K_t(x, x)dx$$

Let $c_1, ..., c_k$ be the positive eigenvalues of P and $c_{k+1}, ..., c_N$ its negative eigenvalues. Then,

$$index(P) = k - (N - k) = 2k - N$$

On the other hand,

$$Tr(exp(-tP)) = \sum_{k=1}^{N} exp(-tc_k)$$

Suppose P is positive semi-definite. Then we can write $P = Q^*Q$ for some operator Q. We consider

$$N(t) = Tr(exp(-tQ^*Q)) - Tr(exp(-tQQ^*))$$

We claim that $N(t) = dim(N(Q)) - dim(N(Q^*)$. This can be proved for example using the singular value decomposition. We write

$$Q = UDV$$

where D is non-negative diagonal and U, V are unitary matrices. Then, $Q^*Q = VD^2V^*, QQ^* = UD^2U^*$ and hence if λ is any non-zero eigenvalue of $P = Q^*Q$, then it is also a non-zero eigenvalue of QQ^* and vice-versa and further the multiplicities of this eigenvalue are the same for both Q^*Q and QQ^*. Thus, the claim follows. Thus for all non-zero t, we have that

$$N(t) = dimN(Q) - dim(N(Q^*)) = dim(N(Q)) - dim(R(Q))^{\perp}$$

[10] **Design of quantum gates using superstring theory** The Lagrangian density for a superstring is given by

$$L = (1/2)h^{\alpha\beta}\sqrt{h}\partial_\alpha X^\mu \partial_\beta X_\mu$$

$$+\psi^{\mu T}\rho^{\alpha}\partial_{\alpha}\psi_{\mu}$$

where $\psi^{\mu}(\tau,\sigma)$ is a Majorana Fermion field and $\rho^{\alpha's}$ are skewsymmetric matrices. It should be noted that for Majorana Fermion fields ψ on $\mathbb{R}^4$, we have

$$\psi^* = \psi^T\gamma_5\epsilon\gamma^0$$

and

$$\bar{\psi} = \psi^*\gamma^0$$

so that

$$\bar{\psi}\gamma^{\mu}\partial_{\mu}\psi = \psi^T\gamma_5\epsilon\gamma^{\mu}\partial_{\mu}\psi$$

We note that $\gamma_5\epsilon\gamma^{\mu}$ and $\epsilon\gamma^{\mu}$ are skew-symmetric matrices and we may as well therefore remove the γ_5 factor to get the Fermionic contribution to the Lagrangian density as

$$\psi^T\epsilon\gamma^{\mu}\partial_{\mu}\psi$$

For strings, however, space-time is two dimensional and hence we should replace $\epsilon\gamma^{\mu}$ by skew-symmetric matrices ρ^{α} to obtain the Fermionic contribution to the Lagrangian density as

$$\psi^T\rho^{\alpha}\partial_{\alpha}\psi$$

The infinitesimal supersymmetry transformations that leave the total superstring action invariant are

$$\delta X^{\mu} = k^T\psi^{\mu}, \delta\psi^{\mu} = \rho^{\alpha T}k\partial_{\alpha}X^{\mu}$$

Then we find that

$$\delta(h^{\alpha\beta}\sqrt{h}X^{\mu}_{,\alpha}X_{\mu,\beta}) =$$

$$= 2h^{\alpha\beta}\sqrt{h}X^{\mu}_{,\alpha}\delta X_{\mu,\beta}$$

$$= 2h^{\alpha\beta}\sqrt{h}X^{\mu}_{,\alpha}k^T\partial_{\beta}\psi_{\mu}$$

and on the other hand,

$$\delta(\psi^{\mu T}\rho^{\alpha}\partial_{\alpha}\psi_{\mu} =$$

$$2\psi^{\mu T}\rho^{\alpha}\partial_{\alpha}\delta\psi_{\mu} =$$

$$2\psi^{\mu T}\rho^{\alpha}\rho^{\beta}kX_{\mu,\alpha\beta}$$

As in the case of the Dirac matrices, we assume that

$$\rho^{\alpha}\rho^{\beta} + \rho^{\beta}\rho^{\alpha} = 2h^{\alpha\beta}$$

and then deduce global supersymmetry invariance of the superstring action integral. To achieve the stronger condition of local supersymmetry, an additional gaugino field contribution has to be added to this superstring action and then local supersymmetry transformations must be appropriately defined.

Exercise: By using a truncated Fourier series expansion of the Bosonic and Fermionic strings in terms of Bosonic and Fermionic creation and annihilation operators acting in the tensor product of the Boson and Fermion Fock space, write down the truncated Hamiltonian for this superstring and then by incorporating interaction terms between the strings and c-number current and potential sources, calculate using the Dyson series, an approximate expression for the unitary evolution operator in terms of the c-number sources and then optimize over these sources so that the unitary evolution operator approximates in the sense of the Frobenius' norm a given unitary gate.

[11] **Project on the design of quantum gates using strings and superstrings**

[1] The superstring Lagrangian density is a quadratic form in the Bosonic and Fermionic component string fields. The corresponding string-field equations give the basic equations for the Bosonic and Fermionic strings in decoupled from. The Bosonic part is simply the wave equation with one time variable and one spatial variable. The Fermionic part is simply the Dirac equation with zero mass in one time and one spatial dimension. The solutions to these equations yield the Bosonic string field as a linear combination of a countably infinite number of Bosonic creation and annihilation operators and the Fermionic string field as a linear combination of a countably infinite of Fermionic creation and annihilation operators. Thus the quantum superstring acts in a tensor product of Boson and Fermion Fock space. The Hamiltonian of the superstring can be expressed as

$$H = \sum_{n\geq 1} c(n)\alpha(-n)\alpha(n) + \sum_{n\geq 1} d(n)\beta(-n)\beta(n)$$

where

$$\alpha(-n) = \alpha(n)^*, \beta(-n) = \beta(n)^*$$

and the $\alpha's$ satisfy the CCR while the $\beta's$ satisfy the CAR:

$$[\alpha(n), \alpha(m)] = \delta(n+m)$$

$$\{\beta(n), \beta(m)\} = \delta(n+m)$$

Now we perturb this Hamiltonian with a gauge field $B_{\mu\nu}(X)$ which corresponds to a string generalization of the electromagnetic potential for particles. The corresponding contribution to the action is given by

$$\int B_{\mu\nu}(X(\tau,\sigma))\epsilon^{\alpha\beta}\partial_\alpha X^\mu \partial_\beta X^\nu d\tau d\sigma$$

It should be noted that the unperturbed action for the superstring is given by

$$S[X,\psi] = (1/2)\int X^\mu_{,\alpha} X^{\mu,\alpha} d\tau d\sigma$$

$$+(1/2)\int \psi^{\mu T}\epsilon.\rho^\alpha \partial_\alpha \psi_\mu d\tau\sigma$$

where $\alpha = 1,2$ and $\alpha = 1$ corresponds to the τ variable while $\alpha = 2$ corresponds to the σ variable. This action has global supersymmetry under Boson-Fermion exchange. This exchange constitutes the global supersymmetry transformation:

$$\delta X^\mu = -k^T \epsilon \psi^\mu$$

$$\delta \psi^\mu = \rho^\alpha k X^\mu_{,\alpha}$$

where k is a two parameter Grassmanian variable. The $\rho^\alpha, \alpha = 1,2$ are 2×2 versions of the Dirac matrices. They satisfy the Dirac anticommutation relations

$$\rho^\alpha \rho^\beta + \rho^\beta \rho^\alpha = 2\eta^{\alpha\beta}$$

where $((\eta^{\alpha\beta})) = diag[1, -1]$ is the string sheet metric. This form of the metric can be obtained by the application of an appropriate Weyl scaling. $\epsilon\rho^\alpha$ are skew-symmetric matrices just as in four space-time dimensions $\epsilon\gamma^\mu$ are skew-symmetric matrices (See S.Weinberg, Vol.III, Supersymmetry). We shall now verify invariance of the above superstring action under the above infinitesimal supersymmetry transformation:

$$\delta((1/2)X^\mu_{,\alpha} X^{;\alpha}_\mu) =$$

$$X^{;\alpha}_\mu \delta X^\mu_{,\alpha} =$$

$$-2X^{;\alpha}_\mu k^T \epsilon \psi^\mu_{,\alpha}$$

$$\delta(\psi^{\mu T} \epsilon \rho^\alpha \psi_{\mu,\alpha}) =$$

$$\delta \psi^{\mu T} \epsilon \rho^\alpha \psi_{\mu,\alpha}$$

$$+\psi^{\mu T} \epsilon \rho^\alpha \delta \psi_{\mu,\alpha}$$

$$= X^\mu_{,\beta} k^T \rho^{\beta T} \epsilon \rho^\alpha \psi_{\mu,\alpha}$$

$$+\psi^{\mu T} \epsilon \rho^\alpha \rho^\beta k X_{\mu,\beta\alpha}$$

$$= -\psi^{\mu T}_{,alpha} \epsilon \rho^\alpha \rho^\beta k X_{\mu,\beta}$$

$$+\psi^{\mu T} \epsilon \rho^\alpha \rho^\beta k X_{\mu,\beta\alpha}$$

$$= -(\psi^{\mu T} \epsilon \rho^\alpha \rho^\beta k X_{\mu,\beta})_{,\alpha}$$

$$+2\psi^{\mu T} \epsilon \rho^\alpha \rho^\beta k X_{\mu,\beta\alpha}$$

$$= -(\psi^{\mu T} \epsilon \rho^\alpha \rho^\beta k X_{\mu,\beta})_{,\alpha}$$

$$+\psi^{\mu T} \epsilon \{\rho^\alpha, \rho^\beta\} k X_{\mu,\beta\alpha}$$

$$= \eta^{\alpha\beta} \psi^{\mu T} \epsilon k X_{\mu,\beta\alpha}$$

where a total two divergence has been neglected since such a term does not contribute to the action integral. Adding the two terms, we find that on neglect of total divergence terms and using the antisymmetry of ϵ, the variation in the action is given by

$$\delta S = -2X^{;\alpha}_\mu k^T \epsilon \psi^\mu_{,\alpha} - X^{;\alpha}_{\mu,\alpha} k^T \epsilon \psi^\mu$$

$$= 0$$

[12] **Superstring action for local supersymmetry transformations**

$$p^\mu_\alpha = (X^\mu_{,\alpha} - \theta^{AT}\epsilon\gamma^\mu\theta^A_{,\alpha})$$

where $A = 1, 2$. Define

$$L_1 = (1/2)p^\mu_\alpha p^\alpha_\mu = (1/2)\eta_{\mu\nu}p^\mu_\alpha p^{\nu\alpha}$$

where α is raised using the worldsheet metric $diag[1, -1]$. Also define

$$L_2 = c_1\epsilon^{\alpha\beta}X^\mu_{,\alpha}(\theta^{1T}\epsilon\gamma_\mu\theta^1_{,\beta} - \theta^{2T}\epsilon\gamma_\mu\theta^2_{,\beta})$$

$$+c_2\epsilon^{\alpha\beta}(\theta^{1T}\gamma^\mu\theta^1_{,\alpha}).(\theta^{2T}\gamma_\mu\theta^2_{,\beta})$$

The local supersymmetry transformations are

$$\delta X^\mu = k^{AT}\epsilon\gamma^\mu\theta^A, \delta\theta^A = k^A$$

where k^A is a Grassmannian parameter dependent on the space-time coordinates τ, σ. We wish to select c_1, c_2 so as to get local supersymmetry of the action $\int(L_1 + L_2 + L_3)d\tau d\sigma$.

Exercise: If need be add additional terms to the superstring action and restrict the dimension of the superstring in order to obtain local supersymmetry. Show that global supersymmetry can be achieved even without putting in these additional terms and without restricting the dimension.

Remark: If gravity has to be included in superstring theory, then necessarily the action must be locally supersymmetric, global supersymmetry is not sufficient.

[13] **Spinors**

Let V be a vector space and Q a quadratic form on V so that for any $u, v \in V$, we have a multiplication $u.v$ satisfying

$$uv + vu = B(u, v)$$

where $B(u, v)$ is the symmetric bilinear form on V induced by Q. Thus,

$$u^2 = B(u, u)/2 = Q(u)$$

and hence

$$B(u, v) = Q(u + v) - Q(u) - Q(v)$$

or equivalently,

$$B(u, v) = (Q(u + v) - Q(u - v))/2$$

We assume that the product $u_1...u_n$ is defined for any $u_1, ..., u_n \in V$. The formal linear span of all such products is denoted by $C(V)$ and $(C(V), Q)$, ie, $C(V)$ equipped with the quadratic form Q is called a Clifford algebra over $C(V)^+$ denotes the subalgebra of $C(V)$ spanned by even number of products. One example of a Clifford algebra is as follows. Let $\wedge$ be an antisymmetric tensor product on V. For $u \in V$, let $a(u)^*$ act on $\wedge V$ by

$$a(u)^* w_1 \wedge ... \wedge u_n = u \wedge w_1 \wedge ... \wedge u_n$$

and let $a(u)$ act on the same by the adjoint operation, ie, contraction:

$$a(u) w_1 \wedge ... \wedge u_n = c(n) \sum_{k=1}^{n} < u, w_k > (-1)^k w_1 ... w_{k-1} \wedge w_{k+1} \wedge ... \wedge w_n$$

Then, it is easy to see that

$$a(u)a(v)^* + a(v)^* a(u) =< u, v >$$

Now we are in a position to describe a construction of a Clifford algebra over an even dimensional vector space. Let V be a real vector space of dimension $2n$ and let $\{e_1, ..., e_{2n}\}$ be a basis for this vector space satisfying

$$(e_k | e_m) = (e_{n+k} | e_{n+m}) = 0, 1 \leq k, m \leq n,$$

$$(e_k | e_{n+m}) = \delta_{km}, 1 \leq k, m \leq n$$

Note that $(.|.)$ is an inner product on V. We wish to construct a Clifford structure so that

$$uv + vu = (u|v), u, v \in V$$

To this end, we consider the n-dimensional real vector space

$$W = span\{e_1, ..., e_n\}$$

and consider the 2^n dimensional vector space $\wedge W$ where $\wedge$ is an antisymmetric tensor product on W. Then define the actions of e_k and $e_{n+k}, 1 \leq k \leq n$ on $\wedge W$ by

$$\gamma(e_k).(u_1 \wedge ... \wedge u_r) = e_k \wedge u_1 \wedge ... \wedge u_r, u_1, ..., u_r \in W$$

and

$$\gamma(e_{n+k})(u_1 \wedge ... \wedge u_r) = c(r) \sum_{m=1}^{r} (-1)^{m-1} (e_{n+k} | u_m)(u_1 \wedge ... \wedge \hat{u}_m \wedge ... \wedge u_r)$$

Then, it is immediate that for $1 \leq k, r \leq n$,

$$\gamma(e_k)\gamma(e_r) + \gamma(e_r)\gamma(e_k) = 0,$$

$$\gamma(e_{n+k})\gamma(e_{n+r}) + \gamma(e_{n+r})\gamma(e_{n+k}) = 0,$$

$$\gamma(e_k)\gamma(e_{n+r})+\gamma(e_{n+r})\gamma(e_k)=\delta(k,r)$$

and hence on extending the map γ linearly and by formally defining an element $u_1...u_r$ with $u_1,...,u_r \in V$ so that

$$\gamma(u_1...u_r)=\gamma(u_1)...\gamma(u_r)$$

we get that

$$\gamma(u)\gamma(v)+\gamma(v)\gamma(u)=(u|v), u, v \in V$$

Thus, we get a Clifford structure on V and it is clear that the dimension of this Clifford algebra is 2^{2n}.

[14] Bosonic string field theory.
The Bosonic string field is given by

$$X^\mu(\tau,\sigma)=x^\mu+p^\mu\tau+i\sum_{n\neq 0}[(\alpha(n)^\mu/n)exp(2\pi in(\tau-\sigma))+(\beta(n)^\mu/n)exp(2\pi in(\tau+\sigma))]$$

where

$$\alpha(n)^{\mu*}=\alpha(-n)^\mu, n\neq 0$$

in order that X^μ be a Hermitian operator field and this string field satisfies the wave equation

$$X^\mu_{,\tau\tau}-X^\mu_{,\sigma\sigma}=0$$

that is derived from the action

$$S[X]=(1/2)\int(X^\mu_{,\tau}X_{\mu,\tau}-X^\mu_{,\sigma}X_{\mu,\sigma})d\tau d\sigma$$

Here, the metric of D-dimensional space-time is given by

$$((\eta_{\mu\nu}))=diag[1,-1,-1,...,-1]$$

The space-time Lorentz group is $SO(1,D-1)$. The canonical momentum field is

$$P_\mu=\delta S/\delta X^\mu_{,\tau}=X_{\mu,\tau}$$

and hence the CCR gives

$$[X^\mu(\tau,\sigma),X_{\nu,\tau}(\tau,\sigma')]=i\delta^\mu_\nu\delta(\sigma-\sigma')$$

or equivalently,

$$[X^\mu(\tau,\sigma),X^\nu_{,\tau}(\tau,\sigma')]=i\eta^{\mu\nu}\delta(\sigma-\sigma')$$

and hence we get

$$[\sum_{n=0}(\alpha(n)^\mu/n)exp(2\pi in(\tau-\sigma))+(\beta(n)^\mu/n)exp(2\pi in(\tau+\sigma)),\sum_{m=0}(\alpha(m)^\mu/m)$$

$$exp(2\pi im(\tau-\sigma))+(\beta(m)^\mu)exp(2\pi im(\tau+\sigma))]=(-\eta^{\mu\nu}/2\pi)\delta(\sigma-\sigma')$$

from which we deduce that

$$[\alpha(n)^\mu, \alpha(m)^\nu] = (-n/4\pi^2)\eta^{\mu\nu}\delta(n+m),$$

$$[\beta(n)^\mu, \beta(m)^\nu] = (-n/4\pi^2)\eta^{\mu\nu}\delta(n+m),$$

$$[\alpha(n)^\mu, \beta(m)^\nu] = 0$$

for all $n, m \neq 0$. We now construct the string field Hamiltonian.

$$H = \int (P_\mu X^\mu_{,\tau} - L)d\sigma =$$

$$= (1/2)\int (X^\mu_{,\tau}X_{\mu,\tau} + X^\mu_{,\sigma}X_{\mu,\sigma})d\sigma$$

$$= p^2/2 - 2\pi^2 \sum_{n\neq 0}(\alpha^\mu(-n)\alpha_\mu(n) + \beta^\mu(-n)\beta_\mu(n))$$

where

$$p^2 = p^\mu p_\mu = p^{02} - p^i p^i$$

The condition that the state of the system have total energy E gives $(H - E)|\phi>= 0$, or formally, on such states, $H = E$. On the other hand, from relativistic mechanics, we know that the mass is given by $M^2 = p^2 = p_\mu p^\mu$. Thus, we get the result that the mass operator of the string is given by

$$2M^2 - 4\pi^2 \sum_{n\geq 1}(\alpha(-n).\alpha(n) + \beta(-n).\beta(n)) = E$$

or equivalently,

$$M^2 = E/2 - 2\pi^2 \sum_{n\geq 1}(\alpha(-n).\alpha(n) + \beta(-n).\beta(n))$$

Formally, the propagator of the (Bosonic) string $\Delta = (H - E)^{-1}$ which can be expressed as

$$\Delta = \int_0^1 z^{H-E-1}dz$$

Equivalently, apart from scaling factors, the Bosonic string propagator is given by

$$\Delta = \int_0^1 z^L dz$$

where

$$L = -a\sum_{n\geq 1}\alpha(n).\alpha(-n)$$

where a is a constant. We are absorbing the other modes $\beta(n)$ into the $\alpha(n)'s$. We can evaluate the matrix elements of the propagator as follows.

$$[\alpha(n)^\mu, \alpha(-n)^\nu] = n\eta^{\mu\nu}$$

and if $|w>$ is a coherent state of the set of harmonic oscillators $\alpha(n), n \geq 1$, we have

$$\alpha(n)|w>=w(n)|w>$$

Suppose $|k_1^{\mu_1}, k_2^{\mu_2}, ...>$ are number states of the system of oscillators. Then, we have

$$\alpha(-n).\alpha(n)|k_1 k_2...>=k_n|k_1 k_2...>$$

where

$$k_n = k_n^0 - k_n^1 - ... - k_n^{D-1}$$

and hence

$$z^{\sum_{n\geq 1}\alpha(-n)\alpha(n)}|k_1 k_2...>=(\Pi_{n\geq 1} z^{k_n})|k_1 k_2...>=z^{\sum_n k_n}|k_1 k_2...>$$

whence

$$<m_1 m_2...|z^L|k_1 k_2...>=z^{-a\sum_n k_n}\delta[\mathbf{m}-\mathbf{k}]$$

[15] **Conformal weights**

Let $A(z)$ be an analytic function of the complex variable z. Suppose that on applying the transformation $w = w(z)$ or equivalently its inverse $z = z(w)$, this function changes to

$$B(w) = A(z)(dz/dw)^J$$

Then we say that A has conformal weight w. Consider an infinitesimal analytic transformation

$$w = z + \epsilon(z)$$

or equivalently,

$$z = w - \epsilon(w)$$

since $\epsilon(z)$ is assumed to be of the first order of smallness. Then, if A has conformal weight J, it gets transformed under this infinitesimal transformation to $A + \delta A$ where

$$A(w) + \delta A(w) = A(z)(1 - \epsilon'(w))^J = A(z)(1 - J\epsilon'(w))$$

or equivalently,

$$A(z) - A'(z)\epsilon(z) + \delta A(z) = A(z)(1 - J\epsilon'(z))$$

so that

$$\delta A(z) = A'(z)\epsilon(z) - JA(z)\epsilon'(z)$$

This is the condition for A to have conformal weight J. Now, consider the following vertex function for a Bosonic string:

$$V(k,z) =: exp(ik.X(z)) := exp(k.\sum_{n\leq -1}\alpha(n)z^n/n).exp(k.\sum_{n\geq 1}\alpha(n)z^n/n)$$

We wish to compute its conformal weight. First we introduce the Fourier components of the energy-momentum tensor:

$$L_n = (1/2)\sum_m \alpha(n-m)\alpha(m)$$

and find that

$$[L_n, X(z)] = (1/2)[\sum_m \alpha(n-m)\alpha(m), i\sum_n (\alpha(n)z^n/n)]$$

$$= (1/2)\sum_m \{\alpha(n-m), [\alpha(m), i\sum_k \alpha(k)z^k/k]\}$$

$$= (1/2)\sum_m \{\alpha(n-m), -iz^{-m}\} = -i\sum_m \alpha(n-m)z^{-m}$$

$$= -iz^{-n}\sum_m \alpha(m)z^m$$

On the other hand,

$$zdX(z)/dz = i\sum_m \alpha(m)z^m$$

and hence we deduce that

$$[L_n, X(z)] = -z^{-n}dX(z)/dz$$

Now consider the infinitesimal transformation $w = z + \delta z = z + \epsilon(z)$. Under this transformation, $X(z)$ changes to $X(z - \epsilon(z)) = X(z) - \epsilon(z)X'(z)$. Hence, we can identify

$$\epsilon(z) = \epsilon.z^{-n}$$

for the infinitesimal Lie transformation $\epsilon.L_n$.

[16] **Super-strings**

$$S[X, \psi] = \int (1/2)\sqrt{h}h^{\alpha\beta}X^{\mu}_{,\alpha}X_{\mu,\beta}d^2\sigma$$

$$-\int \psi^a \epsilon \rho^\alpha \psi^a_{,\alpha}d^2\sigma$$

Weyl scaling: First, we can choose our string coordinates (τ, σ) so that

$$((h_{\alpha\beta})) = exp(\phi)diag[1, -1]$$

and then

$$h = exp(2\phi), ((h^{\alpha\beta})) = exp(-\phi)diag[1, -1], ((\sqrt{h}h^{\alpha\beta})) = diag[1, -1]$$

and so there is no need for Weyl scaling. Already, in our system of coordinates, the Bosonic part of the string action is

$$(1/2)\int \eta^{\alpha\beta} X^{\mu}_{,\alpha} X_{\mu,\beta} d^2\sigma$$

The total angular momentum tensor

$$J^{\mu\nu} = \int (X^{\mu} X^{\nu}_{,\tau} - X^{\nu} X^{\mu}_{,\tau}) d\sigma$$

is conserved. Indeed, by the equations of motion

$$dJ^{\mu\nu}/d\tau = \int_0^1 (X^{\mu} X^{\nu}_{,\tau\tau} - X^{\nu} X^{\mu}_{,\tau\tau}) d\sigma$$

$$= \int_0^1 (X^{\mu} X^{\nu}_{,\sigma\sigma} - X^{\nu} X^{\mu}_{,\sigma\sigma}) d\sigma$$

$$= \int_0^1 (X^{\mu} X^{\nu}_{,\sigma} - X^{\nu} X^{\mu}_{,\sigma})_{,\sigma} d\sigma = 0$$

The conservation of the angular momentum can also be deduced as a Noether current conservation corresponding to the invariance of the Lagrangian under Lorentz transformations of the D-dimensional space-time. Writing

$$X^{\mu}(\tau,\sigma) = i\sum_{n\neq 0}[(\alpha^{\mu}(n)/n)exp(-2\pi in(\tau-\sigma)) + (\beta^{\mu}(n)/n)exp(-2\pi i(\tau+\sigma))]$$

we get that

$$J^{\mu\nu} = \sum_{n\neq 0} n^{-1}(\alpha^{\mu}(n)\alpha^{\nu}(-n) - \alpha^{\nu}(n)\alpha^{\mu}(-n))$$

$$+\sum_{n\neq 0} n^{-1}(\beta^{\mu}(n)\beta^{\nu}(-n) - \beta^{\nu}(n)\beta^{\mu}(-n))$$

Note that we've used

$$[\alpha^{\mu}(n), \beta^{\nu}(m)] = 0$$

Now we look at the Fermionic part of the super-string equations

$$\rho^{\alpha}\psi^{a}_{,\alpha} = 0$$

ie,

$$\rho^{0}\psi^{a}_{,\tau} + \rho^{1}\psi^{a}_{,\sigma} = 0$$

We first start with the Fermionic Lagrangian density

$$L_F = \psi^{a}\epsilon\rho^{\alpha}\psi^{a}_{,\alpha}$$

$$= \psi^{a}\epsilon\rho^{0}\psi^{a}_{,\tau} + \psi^{a}\epsilon\rho^{1}\psi^{a}_{,\sigma}$$

The canonical momentum field is

$$P_a = \partial L_F / \partial \psi^a_{,\tau} = -\epsilon \rho^0 \psi^a$$

Note that $\epsilon \rho^\alpha, \alpha = 0, 1$ are 2×2 skew-symmetric matrices. Thus, the Hamiltonian density is given by

$$\psi^a = \rho^0 \epsilon P_a$$

and

$$H_F = P_a^T \psi^a_{,\tau} - L_F = -\psi^a \epsilon \rho^1 \psi^a_{,\sigma}$$

The CAR are

$$\{\psi^a(\tau,\sigma), P_b(\tau,\sigma')^T\} = \delta^a_b \delta(\sigma - \sigma') I_2$$

or equivalently,

$$\{\psi^a(\tau,\sigma), \psi^b(\tau,\sigma')^T\} \epsilon \rho^0 = \delta^a_b \delta(\sigma - \sigma') I_2$$

or equivalently,

$$\{\psi^a(\tau,\sigma), \psi_b(\tau,\sigma')^T\} = -\delta^a_b \delta(\sigma - \sigma') \rho^0 \epsilon$$

Note that ψ^a are Majorana Fermions, which means that

$$\psi^{a*} = \psi^{aT} \epsilon \rho^0$$

and hence

$$\bar{\psi}^a = \psi^{a*} \rho^0 = \psi^{aT} \epsilon$$

We can expand the solution as

$$\psi^a(\tau,\sigma) = \sum_n S^a(n,\tau) exp(-2\pi i n \sigma)$$

For this to satisfy the above equation of motion, we require that

$$\rho^0 \partial_\tau S^a(n,\tau) = 2\pi i n \rho^1 S^a(n,\tau)$$

and hence

$$S^a(n,\tau) = exp(2\pi i n \tau A) S^a(n)$$

where

$$A = \rho^0 \rho^1$$

The eigenvalues of A are ± 1 and therefore as in the Bosonic case, we again get the result that in the Fermionic case, the wave field is a superposition of a forward and a backward travelling wave:

$$A = e_0 e_0^T - e_1 e_1^T, e_0^T e_1 = 0, e_0^T e_0 = e_1^T e_1 = 1$$

and hence

$$exp(2\pi i n \tau A) = exp(2\pi i n \tau) e_0 e_0^T + exp(-2\pi i n \tau) e_1 e_1^T$$

and hence

$$S^a(n,\tau) = exp(2\pi in\tau)P_0S^a(n) + exp(-2\pi in\tau)P_1S^a(n)$$

where

$$P_0 = e_0e_0^T, P_1 = e_1e_1^T$$

and hence,

$$\psi^a(\tau,\sigma) = \sum_n [P_0S^a(n)exp(2\pi in(\tau-\sigma)) + P_1S^a(n)exp(-2\pi in(\tau+\sigma))]$$

For simplicity of notation, we denote $P_0S^a(n)$ by $S^a(n)$ and $P_1S^a(n)$ by $T^a(n)$. Then, we can express the solution for this two dimensional massless Dirac equation as

$$\psi^a(\tau,\sigma) = \sum_n S^a(n)exp(2\pi in(\tau-\sigma)) + T^a(n)exp(-2\pi in(\tau+\sigma))$$

We are now in a position, to derive the CAR satisfied by $S^a(n), T^a(n), n \in \mathbb{Z}$ for our Fermionic string. Take this as an exercise.

[17] **Energy-Momentum tensor and super-current for a superstring**

$$L = (1/2)X^{\mu}_{,\alpha}X_{\mu}^{;\alpha} + \psi^{\mu T}\rho^0(\rho^0\partial_0 + \rho^1\partial_1)\psi_{\mu}$$

where

$$\rho^0 = \sigma_1, \rho^1 = i\sigma_2$$

Thus

$$(\rho^0)^2 = I, \rho^0\rho^1 = i\sigma_1\sigma_2 = -\sigma_3$$

Then writing

$$\psi^{\mu} = [\psi^{\mu}_+, \psi^{\mu}_-]^T$$

we get that the Fermionic component of the superstring Lagrangian density is given by

$$L_F = \psi^{\mu T}\rho^0(\rho^0\partial_0 + \rho^1\partial_1)\psi_{\mu}$$

$$= \psi^{\mu T}(I\partial_0 - \sigma_3\partial_1)\psi_{\mu} =$$

$$\psi^{\mu}_+\partial_-\psi_{\mu+} + \psi^{\mu}_-\partial_+\psi_{\mu-}$$

where

$$\partial_{\pm} = \partial_0 \pm \partial_1$$

Note that we may define

$$x_+ = (\tau+\sigma)/2, x_- = (\tau-\sigma)/2$$

and then

$$\partial_0 = (1/2)(\partial/\partial x_+ + \partial/\partial x_-)$$

$$\partial_1 = (1/2)(\partial/\partial x_+ - \partial/\partial x_-)$$

and hence

$$\partial/\partial x_+ = \partial_0 + \partial_1 = \partial_+,$$

$$\partial/\partial x_- = \partial_0 - \partial_1 = \partial_-$$

The Euler-Lagrange equations for the Fermionic components are easily seen to be

$$\partial_+\psi_- = 0, \partial_-\psi_+ = 0$$

where ψ_+ is an abbreviation for $((\psi_+^\mu))$ and likewise ψ_- is an abbreviation for ψ_-^μ. These equations imply that ψ_+ is any function of $\tau - \sigma$ only and ψ_- is any function of $\tau + \sigma$ only. The solutions to these equations are of two kinds depending upon the boundary conditions. These boundary conditions can be described as follows. The spatial string variable σ is assumed to vary over $[0, \pi]$. The variation in $S_F = \int_0^\pi L_F d\sigma$ gives us the following boundary term on integration by parts:

$$\psi_+\delta\psi_+ - \psi_-\delta\psi_-$$

This must vanish at the boundary, ie, at $\sigma = 0, \pi$. To make it vanish at $\sigma = 0$, we assume the boundary condition that

$$\psi_+(\tau, 0) = \psi_-(\tau, 0)$$

and hence

$$\delta\psi_+(\tau, 0) = \delta\psi_-(\tau, 0)$$

To make it vanish at $\sigma = \pi$, we may assume that the boundary condition that either

$$\psi_+(\tau, \pi) = \psi_-(\tau, \pi)$$

and hence

$$\delta\psi_+(\tau, \pi) = \delta\psi_-(\tau, \pi)$$

or else

$$\psi_+(\tau, \pi) = -\psi_-(\tau, \pi)$$

and hence

$$\delta\psi_+(\tau, \pi) = -\delta\psi_-(\tau, \pi)$$

The first kind of boundary conditions implies that $\psi_\pm$ admit the modal expansions

$$\psi_+^\mu(\tau, \sigma) = \sum_{n\in\mathbb{Z}} b^\mu(n)exp(in(\tau - \sigma)),$$

$$\psi_-^\mu(\tau, \sigma) = \sum_{n\in\mathbb{Z}} b^\mu(n)exp(in(\tau + \sigma))$$

and the second kind of boundary conditions implies that

$$\psi_+^\mu(\tau, \sigma) = \sum_{n\in\mathbb{Z}+1/2} b^\mu(n)exp(in(\tau - \sigma)),$$

$$\psi^\mu_-(\tau,\sigma) = \sum_{n\in\mathbb{Z}+1/2} b^\mu(n)exp(in(\tau+\sigma))$$

This is the situation for general strings, open or closed. If we further restrict to closed strings, then we must impose the periodic boundary conditions, that for any solutions, the value at $\sigma = 0$ must coincide with the value at $\sigma = \pi$. This additional restriction implies that we obtain the following modal expansions:

$$\psi^\mu_+(\tau,\sigma) = \sum_{n\in\mathbb{Z}} b^\mu(n)exp(2in(\tau-\sigma))$$

and

$$\psi^\mu_-(\tau,\sigma) = \sum_{n\in\mathbb{Z}} b^\mu(n).exp(2in(\tau+\sigma))$$

for the first kind of boundary conditions and for the second kind, we do not have any periodic solutions unless we extend the spatial domain to the range $[-\pi,\pi]$ in which case, we get the (2π periodic) solution

$$\psi^\mu_+(\tau,\sigma) = \sum_{n\in\mathbb{Z}+1/2} b^\mu(n)exp(2in(\tau-\sigma))$$

and

$$\psi^\mu_-(\tau,\sigma) = \sum_{n\in\mathbb{Z}+1/2} b^\mu(n).exp(2in(\tau+\sigma))$$

We now compute the components of the energy-momentum tensor as well as the super-current of our superstring. The Fermionic components are

$$T^+_{F+} = (\partial L_F/\partial\partial_+\psi_-).\partial_+\psi_- =$$

$$\psi_-.\partial_+\psi_- = 0$$

$$T^-_{F+} = (\partial L_F/\partial\partial_-\psi_+).\partial_+\psi_+$$

$$= \psi_+.\partial_+\psi_+$$

$$T^+_{F-} = (\partial L_F/\partial\partial_+\psi_-).\partial_-\psi_-$$

$$= \psi_-.\partial_-\psi_-$$

and finally,

$$T^-_{F-} = (\partial L_F/\partial\partial_-\psi_+).\partial_-\psi_+$$

$$\psi_+\partial_-\psi_+ = 0$$

Now, we choose a coordinate system followed by a Weyl rescaling so that the metric becomes

$$ds^2 = d\tau^2 - d\sigma^2 = d(\tau-\sigma).d(\tau+\sigma) = dx_-dx_+$$

so that in the $(x+, x_-)$ system, the world-sheet metric tensor is given by

$$\begin{pmatrix} g_{++} & g_{+-} \\ g_{-+} & g_{--} \end{pmatrix} =$$

$$\begin{pmatrix} 0 & 1 \\ 1 & 0 \end{pmatrix}$$

ie,

$$g_{++} = g_{--} = 0, g_{+-} = g_{-+} = 1$$

and then we get the covariant components of the Fermionic part of the energy-momentum tensor are given by

$$T_{F++} = g_{+-} T^{-}_{F+} = T^{-}_{F+} = \psi_{+}.\partial_{+}\psi_{+},$$

Likewise,

$$T_{F--} = g_{-+} T^{+}_{F-} = \psi_{-}\partial_{-}\psi_{-}$$

$$T_{F+-} = g_{+-} T^{-}_{F-} = T^{-}_{F-} = \psi_{+}\partial_{-}\psi_{+} = 0$$

$$T_{F-+} == g_{-+} T^{+}_{F+} = 0$$

We note the Fermionic energy-momentum conservation laws

$$\partial^{+} = g^{+-}\partial_{-} = \partial_{-},$$

$$\partial^{-} = g^{-+}\partial_{+} = \partial_{+}$$

and hence

$$\partial^{+} T_{F++} + \partial^{-} T_{F-+} = \partial^{+} T_{FF_{+}+} = \partial_{-} T_{F++} = \partial_{-}(\psi_{+}\partial_{+}\psi_{+}) = 0$$

since

$$\partial_{-}\psi_{+} = 0, \partial_{-}\partial_{+} = \partial_{+}\partial_{-}$$

Likewise,

$$\partial^{-} T_{F--} + \partial^{+} T_{F+-} = \partial_{+} T_{F--} = \partial_{+}(\psi_{-}\partial_{+}\psi_{-}) = 0$$

since

$$\partial_{+}\psi_{-} = 0$$

We now also observe that the Noether theorem applied to the supersymmetry invariance of the combined Boson-Fermion action implies the conservation of the supercurrent. The components of the supercurrent are obtained by observing that the variation of the total action under an infinitesimal local supersymmetry transformation is given by and expression of the form

$$\delta_{\epsilon,susy} S = \int (\partial_{\alpha}\epsilon(\sigma)(\sigma)) J^{\alpha}(\sigma) d^{2}\sigma$$

and hence if the equations of motion are satisfied, then the above variation must vanish for all infinitesimal local parameters ϵ and hence the supercurrent must be conserved:

$$\partial_{\alpha} J^{\alpha}(\sigma) = 0$$

This conservation law can be stated in an alternate form as

$$\partial_- J_+ = 0, \partial_+ J_- = 0$$

where

$$J_+ = \partial_+ X^\mu \psi_{\mu+}, J_- = \partial_- X^\mu \psi_{\mu-}$$

In this form, it is immediate to see these currents are conserved from the equations of motion:

$$\partial_- \partial_+ = \partial^\alpha \partial_\alpha$$

and hence

$$\partial_- \partial_+ X^\mu = 0,$$

$$\partial_- \psi_{\mu+} = 0$$

[18] **Super-symmetric, gauge invariant and Lorentz invariant action for non-Abelian gauge fields**

$$D = \gamma_5 \epsilon \partial_\theta - \gamma^\mu \theta \partial_\mu$$

$$D_L = (1+\gamma_5)D/2, D_R = (1-\gamma_5)D/2$$

$V^A(x,\theta)$ is the gauge superfield which in the Wess-Zumino gauge, can be expressed as

$$V^A(x,\theta) = \theta^T \epsilon \gamma^\mu \theta . V_\mu^A(x)$$

$$+\theta^T \epsilon \theta . \theta^T \epsilon \lambda^A(x) + (\theta^T \epsilon \theta)^2 D^A$$

We put

$$\theta_L = (1+\gamma_5)\theta/2, \theta_R = (1-\gamma_5)\theta/2$$

Then,

$$D_L = \epsilon \partial_{\theta_L} - \gamma^\mu \theta_R \partial_\mu$$

$$D_R = -\epsilon \partial_{\theta_R} - \gamma^\mu \theta_L \partial_\mu$$

where we have used the identity,

$$\gamma_5 \gamma^\mu + \gamma^\mu \gamma_5 = 0,$$

Note that V^A is not a Chiral field. We write

$$t.V = t_A V^A$$

where summation over the non-Abelian gauge index A is implied. $\{t_A\}$ form a complete set of Hermitian generators for the gauge group assumed to be a subgroup of $U(N)$. We define the left Chiral fields

$$W_a^A(x,\theta) = D_R^T \epsilon . D_R(exp(-t.V).D_{L_a} exp(t.V))$$

Note that Φ is a left Chiral superfield iff $D_R\Phi = 0$ and it is right Chiral iff $D_L\Phi = 0$. By definition, Φ is left Chiral iff it is a function of only θ_L and

$$x_+^\mu = x^\mu + \theta_R^T\epsilon\gamma^\mu\theta_L$$

Note that

$$D_R x_+^\mu = -\gamma^\mu\theta_L + (-\epsilon)(\epsilon\gamma^\mu\theta_L) = 0$$

and likewise, Φ is right Chiral iff it is a function of only θ_R and

$$x_-^\mu = x^\mu - \theta_R^T\epsilon\gamma^\mu\theta_L$$

We note that

$$D_L x_-^\mu = 0$$

Also,

$$D_R\theta_{La} = -\epsilon\partial_{\theta_R}\theta_{La} = 0$$

and likewise,

$$D_L\theta_{Ra} = 0$$

since

$$(1+\gamma_5)(1-\gamma_5) = (1-\gamma_5)(1+\gamma_5) = 0$$

These relations can be expressed in matrix notation as

$$D_L\theta_R^T = D_R\theta_L^T = 0$$

Also note that

$$\theta_R^T\epsilon\gamma^\mu\theta_L = \theta^T(1-\gamma_5)\epsilon\gamma^\mu(1+\gamma_5)\theta/4$$
$$= \theta^T(1-\gamma_5)\epsilon\gamma^\mu\theta/2$$

Then,

$$V^A(x,\theta) = \theta^T\epsilon\gamma^\mu\theta.V_\mu^A(x)$$
$$+\theta^T\epsilon\theta.\theta^T\epsilon\lambda^A(x) + (\theta^T\epsilon\theta)^2 D^A$$

gives

$$exp(t.V) = 1 + \theta^T\epsilon\gamma^\mu\theta.V_\mu^A(x)t_A$$
$$+\theta^T\epsilon\theta.\theta^T\epsilon\lambda^A(x)t_A + (\theta^T\epsilon\theta)^2 D^A t_A$$
$$+(\theta^T\epsilon\gamma^\mu\theta.V_\mu^A(x)t_A)^2/2$$

The last term here is the same as

$$\theta^T\epsilon\gamma^\mu\theta.\theta^T\gamma^\nu\theta V_\mu^A V_\nu^B t_A t_B$$

[19] **Lagrangian for Abelian gauge superfields**
[1] Show that

$$L = c_1 F_{\mu\nu}F^{\mu\nu} + c_2\lambda^T\gamma_5\epsilon\gamma^\mu\partial_\mu\lambda$$

$$+c_3D^2$$

is supersymmetry invariant for an appropriate choice of the constants c_1, c_2, c_3. Here

$$F_{\mu\nu} = V_{\nu,\mu} - V_{\mu,\nu}$$

Here the superfield is given by

$$S[x,\theta] = C(x) + \theta^T\epsilon\omega(x) + \theta^T\epsilon\theta M(x)$$
$$+\theta^T\gamma_5\epsilon\theta N(x) + \theta^T\epsilon\gamma^\mu\theta.V_\mu(x)$$
$$+\theta^T\epsilon\theta\theta^T\gamma_5\epsilon(\lambda(x) + a.\gamma^\mu\partial_\mu\omega(x))$$
$$+(\theta^T\epsilon\theta)^2(D(x) + b\Box C(x))$$

The supersymmetry generator is given by α^TL where α is a Majorana Fermionic parameter and

$$L = \gamma_5\epsilon\partial_\theta + \gamma^\mu\theta\partial_\mu$$

We have

$$\delta C(x) = \alpha^T\gamma_5\epsilon^2\omega(x) = -\alpha^T\gamma_5\omega(x),$$
$$\theta^T\epsilon\theta\delta M(x)$$
$$+\theta^T\gamma_5\epsilon\theta\delta N(x) + \theta^T\epsilon\gamma^\mu\theta.\delta V_\mu(x)$$
$$= \alpha^T\gamma^\mu\theta\theta^T\epsilon\omega_{,\mu}(x)$$
$$+\alpha^T\gamma_5\epsilon\partial_\theta[\theta^T\epsilon\theta.\theta^T\gamma_5\epsilon(\lambda + a\gamma^\mu\omega_{,\mu})]$$

Now, we observe that for any function $w(x) \in \mathbb{C}^4$, we have

$$\partial_\theta(\theta^T\epsilon\theta.\theta^Tw)$$
$$= 2\epsilon\theta.\theta^Tw + \theta^T\epsilon\theta.w$$

[20] Problems in supersymmetric quantum theory.
[1] Consider the superfield

$$S[x,\theta] = C(x) + \theta^T\epsilon\omega(x) + \theta^T\epsilon\theta.M(x)$$
$$+\theta^T\gamma_5\epsilon\theta.N(x) + \theta^T\epsilon\gamma^\mu\theta.V_\mu(x)$$
$$+\theta^T\epsilon\theta.\theta^T\gamma_5\epsilon(\lambda(x) + a.\gamma^\mu\omega_{,\mu})$$
$$+(\theta^T\epsilon\theta)^2(D(x) + b.\Box C(x))$$

The infinitesimal supersymmetry transformation is

$$D = \alpha^T\gamma_5\epsilon L$$

where

$$L = \gamma_5\epsilon\partial_\theta + \gamma^\mu\theta\partial_\mu$$

Then the change in the component fields under such an infinitesimal supersymmetry transformation are given by

$$\delta C(x) = -\gamma_5 \alpha \omega_{,\mu}(x)$$

$$\theta^T \epsilon \delta\omega(x) = \alpha^T \gamma_5 \epsilon.[\gamma^\mu \theta C_{,\mu}(x) + \gamma_5 \epsilon.2\epsilon\theta M + 2(\gamma_5\epsilon)^2 \theta N + 2\gamma_5 \epsilon^2 \gamma^\mu \theta.V_\mu]$$

or equivalently,

$$\epsilon\delta\omega(x) = (\gamma_5 \epsilon \gamma^\mu)^T \alpha.C_{,\mu} + 2\epsilon\alpha M$$
$$+2\gamma_5 \epsilon\alpha.N + 2\epsilon\gamma^\mu \alpha V_\mu$$
$$= \gamma_5 \epsilon \gamma^\mu \alpha.C_{,\mu} + 2\epsilon\alpha M$$
$$+2\gamma_5 \epsilon\alpha.N + 2\epsilon\gamma^\mu \alpha V_\mu$$

or equivalently,

$$\delta\omega(x) = \gamma_5 \gamma^\mu \alpha.C_{,\mu} + 2\alpha.M + 2\gamma_5 \alpha.N + 2\gamma^\mu \alpha.V_\mu$$

Likewise,

$$\theta^T \epsilon\theta.\delta M + \theta^T \gamma_5 \epsilon\theta.\delta N + \theta^T \epsilon\gamma^\mu \theta.\delta V_\mu$$
$$= \alpha^T \gamma_5 \epsilon[\gamma^\mu \theta.\theta^T \epsilon\omega_{,\mu}$$
$$+\gamma_5 \epsilon \partial_\theta (\theta^T \epsilon\theta\theta^T \gamma_5 \epsilon(\lambda + a\gamma^\mu \omega_{,\mu}))]$$
$$= \alpha^T \gamma_5 \epsilon[\gamma^\mu \theta.\theta^T \epsilon\omega_{,\mu}$$
$$+\gamma_5 \epsilon(2\epsilon\theta\theta^T \gamma_5 \epsilon(\lambda + a\gamma^\mu \omega_{,\mu}) + (\theta^T \epsilon\theta)\gamma_5 \epsilon(\lambda + a\gamma^\mu \omega_{,\mu})]$$
$$= \alpha^T \gamma_5 \epsilon\gamma^\mu \theta.\theta^T \epsilon\omega_{,\mu}$$
$$-2\alpha^T \epsilon\theta\theta^T \gamma_5 \epsilon(\lambda + a\gamma^\mu \omega_{,\mu}) - (\theta^T \epsilon\theta)\alpha^T \gamma_5 \epsilon(\lambda + a\gamma^\mu \omega_{,\mu})$$

Writing

$$\theta\theta^T = c_1 \theta^T \epsilon\theta\epsilon + c_2 \theta^T \gamma_5 \epsilon\theta.\gamma_5 \epsilon + c_3 \theta^T \epsilon\gamma^\mu \theta.\epsilon\gamma_\mu$$

we get on equating coefficients of $\theta^T \epsilon\theta$, $\theta^T \gamma_5 \epsilon\theta$ and $\theta^T \epsilon\gamma^\mu \theta$ respectively, the equations

$$\delta M = \alpha^T \gamma_5 \epsilon(2c_1 \lambda + ((2c_1 - 1)a - c_1)\gamma^\mu \omega_{,\mu}),$$
$$\delta N = 2c_2 \alpha^T \epsilon\lambda + c_2(1 + 2a)\alpha^T \epsilon\gamma^\mu \omega_{,\mu}$$
$$= c_2 \alpha^T \epsilon(2\lambda + (1 + 2a)\gamma^\mu \omega_{,\mu})$$

Exercise: Prove that if V_μ is a pure $U(1)$ gauge ie, $V_\mu = \partial_\mu Z$ or equivalently if $V_{\nu,\mu} - V_{\mu,\nu} = 0$ and further if $D = 0, \lambda = 0$, then the form of the above superfield S remains invariant under supersymmetry transformations. Such superfield describe Chiral matter fields and they enable us to construct matter Lagrangians in contrast to gauge Lagrangians for which once the Wess-Zumino gauge has been adopted so that only the λ, D, V_μ components are nonzero, then one can construct supersymmetric gauge Lagrangians from these that are invariant under ordinary gauge transformations, Loretnz transformations and in addition have supersymetry when the left Chiral gauge spinor superfield W_L is constructed appropriately from λ, D, V_μ.

[21] **Bosonic string theory**: Derivation of the Einstein field equations for gravitation in vacuum based on conformal invariance of the string action.

(τ, σ) represent respectively the time variable and the length variable along the string, $\tau \geq 0, 0 \leq \sigma \leq 1$. The string is assumed to be D dimensional, so that any point on its surface is parameterized by (τ, σ): $X^\mu = X^\mu(\tau, \sigma)$. The space-time metric on the string surface is a two dimensional metric given by $h_{\alpha\beta}(\tau, \sigma)$. Thus, the string action functional is given by

$$S_1(X) = \int h^{\alpha\beta}(\tau, \sigma)\sqrt{-h} g_{\mu\nu}(X(\tau, \sigma)) X^\mu_{,\alpha} X^\nu_{,\beta} d\tau d\sigma$$

We usually assume the string metric to be flat, ie

$$((h^{\alpha\beta})) = diag[1, -1]$$

Then writing

$$\partial^\alpha = h^{\alpha\beta}\partial_\beta$$

we get

$$\partial^0 = \partial_0 = \partial/\partial\tau,$$
$$\partial^1 = -\partial_1 = -\partial/\partial\sigma$$

and then the string action can be expressed as

$$S_1[X] = \int g_{\mu\nu}(X)\partial_\alpha X^\mu . \partial^\alpha X^\nu d\tau d\sigma$$

Under an infinitesimal conformal Weyl transformation, the metric changes to

$$exp(\epsilon\phi(X)) g_{\mu\nu}(X) = (1 + \epsilon.\phi(X)).g_{\mu\nu}(X)$$

and then the change in the string action becomes

$$\epsilon \int \phi(X).g_{\mu\nu}(X).\partial_\alpha X^\mu . \partial^\alpha X^\nu d\tau . d\sigma$$

What we require is a quantum average of this variation in the action. To evaluate this, we write

$$X^\mu(\tau, \sigma) = X_0^\mu(\tau, \sigma) + x^\mu(\tau, \sigma)$$

where $x^\mu(\tau, \sigma)$ is a small quantum fluctuation. The propagator of this fluctuation can be derived using the Green's function method applied to the equations of motion of the free string. Specifically, we have the flat space-time equations of motion

$$\partial_\tau^2 X^\mu(\tau, \sigma) = \partial_\sigma^2 X^\mu(\tau, \sigma)$$

and then defining the propagator as

$$G^{\mu\nu}(\tau, \sigma|\tau', \sigma') = < T(X^\mu(\tau, \sigma).X^\nu(\tau', \sigma')) >$$

$$= \theta(\tau - \tau') < X^\mu(\tau, \sigma).X^\nu(\tau', \sigma') > +\theta(\tau' - \tau) < X^\nu(\tau', \sigma').X^\mu(\tau, \sigma) >$$

the equations of motion and the equal time Bosonic commutation relations

$$[\partial_\tau X^\mu(\tau, \sigma), X\nu(\tau, \sigma')] = \eta^{\mu\nu}\delta(\sigma - \sigma')$$

imply that the string propagator satisfies the following pde:

$$(\partial_\tau^2 - \partial_\sigma^2)G^{\mu\nu}(\tau, \sigma|\tau', \sigma') = \eta^{\mu\nu}\delta(\tau - \tau').\delta(\sigma - \sigma')$$

Hence we can formally express this propagator as

$$G^{\mu\nu}(\tau, \sigma|\tau', \sigma') = G^{\mu\nu}(\tau - \tau', \sigma - \sigma') =$$

$$\eta^{\mu\nu} \int \frac{d^2k}{k^2} exp(i(k_1(\tau - \tau') - k_2(\sigma - \sigma')))$$

In this expression,

$$d^2k = dk_1 dk_2, k^2 = k_1^2 - k_2^2$$

We now evaluate the change in the string action caused by a Weyl conformal transformation of the metric:

$$g_{\mu\nu}(X) \to exp(\epsilon\phi(X))g_{\mu\nu}(X) = (1 + \epsilon\phi(x))g_{\mu\nu}(X)$$

The change in the action under such a transformation is

$$\delta S[X] = \epsilon \int \phi(X)g_{\mu\nu}(X)\partial_\alpha X^\mu.\partial^\alpha X^\nu d\tau d\sigma$$

Now we evaluate the average of this quantity:

$$g_{\mu\nu}(X) \approx g_{\mu\nu}(X_0) + (1/2)g_{\mu\nu,\rho\sigma}(X_0)x^\rho x^\sigma$$

In a system of normal coordinates around $X_0^\mu(\tau, \sigma)$, we have

$$g_{\mu\nu,\rho\sigma}(X_0) = R_{\mu\rho\nu\sigma}(X_0)$$

since in such a normal coordinate system, the first order partial derivatives of $g_{\mu\nu}$ vanish at X_0. Further, from the above calculation of the propagator, using dimensional regularization,

$$< x^\rho(\tau, \sigma).x^\sigma(\tau, \sigma) >= \eta^{\rho\sigma} lim_{\tau\to\tau', \sigma'\to\sigma}$$

$$\int d^{2+\epsilon}k.exp(i(k_1(\tau-\tau')-k_2(\sigma-\sigma')))d^2k/k^2$$

$$= \eta^{\rho\sigma} \int k^{1+\epsilon}.dk/k^2 = \int dk/k^{1-\epsilon} \approx \eta^{\rho\sigma}/\epsilon$$

and hence we get

$$< \delta S[X] >= \eta^{\rho\sigma} \int \phi(X_0)R_{\mu\rho\nu\sigma}(X_0(\tau, \sigma))\partial_\alpha X_0^\mu \partial^\alpha X_0^\nu d\tau d\sigma$$

$$= \int \phi(X_0)R_{\mu\nu}(X_0)\partial_\alpha X_0^\mu \partial^\alpha X_0^\nu d\tau d\sigma$$

and the condition for this variation to be zero, ie conformal invariance of the quantum averaged string action is that

$$R_{\mu\nu} = 0$$

ie the Einstein field equations be satisfied. This is true in a normal coordinate system, but since $R_{\mu\nu}$ is a tensor, it should be true in all reference frames. Thus, we have proved that the Einstein field equations in vacuum naturally follow from the conformal invariance of the string action.

[22] **Virasoro algebra in bosonic quantum string theory**
Study project. Expand the bosonic string field as a Fourier series describing forward and backward propagating waves having as coefficients creation and annihilation operators and express the components of the energy momentum tensor as a Fourier series with coefficients being quadratic combinations of creation and annihilation operators. Derive the Virasoro Lie algebra commutation relations between these Fourier coefficients of the energy-momentum tensor introducing central charge terms caused by the ambiguity of the operator ordering in the energy component.

A remark from Edward Witten's lecture: Canonical Relativistic quantum field theory makes gravity an impossibility (due to renormalization problems) while string theory makes gravity inevitable (due to the appearance of the Einstein field equations when conformal invariance of the action is imposed).

[23] **Some aspects of Quantum superstring theory**
[1] Computation of the Veneziano amplitude. Let Δ denote the Bosonic string propagator. It can be represented as

$$\Delta = (L_0 - 1)^{-1}, L_0 = p^2/2 + (1/2)\sum_n \alpha(-n).\alpha(n) = (1/2)(p^2 + N)$$

A physical state $|\psi>$ is defined by the following conditions that it must satisfy:

$$(L_0 - a)|\psi >= 0, L_m|\psi >= 0, m > 0$$

where

$$L_m = (1/2)\sum_n \alpha(m-n).\alpha(n), m \in \mathbb{Z}$$

What is the significance of the operators L_m?. These operators arise while as Fourier coefficients of the string energy-momentum tensor and in classical general relativity of a single particle we have the condition that the velocity four vector of a single point particle has unit length:

$$\sqrt{g_{\mu\nu}(x)v^\mu v^\nu} = 1, v^\mu = dx^\mu/d\tau$$

The generalization of this condition to the quantum theory in the case of strings is that a physical state should be annihilated by the positive frequency components of the energy-momentum tensor $L_m, m > 0$. The super-string Lagrangian is given by

$$L(X, \psi) = (1/2)\partial_\alpha X^\mu.\partial^\alpha X_\mu - i\bar{\psi}\rho^\alpha\partial_\alpha\psi$$

where

$$\rho^0 = \sigma_2, \rho^1 = i\sigma_1, \epsilon_0 = i\sigma_2$$

and

$$\bar{\psi} = \psi^T\epsilon_0$$

so that

$$-i\bar{\psi}\rho^\alpha\partial_\alpha\psi =$$

$$-i\psi^T\epsilon_0\rho^\alpha\partial_\alpha\psi$$

Now,

$$-i\epsilon_0\rho^0 = \sigma_2.\sigma_2 = I_2,$$

$$-i\epsilon_0\rho^1 = \sigma_2.i\sigma_1 = \sigma_3$$

Thus, writing

$$\psi = \begin{pmatrix} \psi_- \\ \psi_+ \end{pmatrix}$$

we get the result that the Fermionic part of the string Lagrangian is given by

$$L_F = \psi_-(\partial_\tau + \partial_\sigma)\psi_- - \psi_+(\partial_\tau - \partial_\sigma)\psi_+$$

$$= \psi_-\partial_+\psi_- + \psi_+\partial_-\psi_+$$

whence the Fermionic field equations are

$$\partial_+\psi_- = 0, \partial_+\psi_- = 0$$

Further, the Bosonic part of the Lagrangian can be expressed as

$$L_B = (1/2)[\partial_\tau X^\mu\partial_\tau X_\mu - \partial_\sigma X^\mu.\partial_\sigma X_\mu)$$

$$= (1/2)\partial_+ X^\mu\partial_- X_\mu$$

The energy-momentum tensor of this superstring field is conserved since this action is invariant under global spatio-temporal translations, ie, under $\tau \to \tau + a$ and $\sigma \to \sigma + b$. This tensor is given by

$$T^\alpha_\beta = (\partial L/\partial X^\mu_{,\alpha})).(X^\mu_{,\beta}$$

$$+(\partial/\partial\partial_\alpha\psi_-))(\partial_\beta\psi_-)$$

$$+(\partial/\partial\partial_\alpha\psi_+))(\partial_\beta\psi_+)$$

where $\alpha, \beta = 0, 1$. If we use the $\sigma^{\pm}$ coordinates in place of τ, σ, then the components of the energy-momentum tensor are

$$T_+^+ = (\partial L/\partial \partial_+ X^\mu).(\partial_+ X^\mu)$$
$$+(\partial/\partial \partial_+ \psi_-))(\partial_+ \psi_-)$$
$$+(\partial/\partial \partial_+ \psi_+))(\partial_+ \psi_+)$$
$$= (1/2)\partial_- X_\mu.\partial_+ X^\mu + \psi_- \partial_+ \psi_-$$

which in the case when the field equations are satisfied reduces to a purely Bosonic component:

$$T_+^+ = (1/2)\partial_- X_\mu.\partial_+ X^\mu$$

Likewise,

$$T_-^- = (\partial L/\partial \partial_- X^\mu).(\partial_- X^\mu)$$
$$+(\partial/\partial \partial_- \psi_-))(\partial_- \psi_-)$$
$$+(\partial/\partial \partial_- \psi_+))(\partial_- \psi_+)$$
$$= (1/2)\partial_- X_\mu.\partial_+ X^\mu + \psi_+ \partial_- \psi_+$$

which once again, in the case when the field equations are satisfied, reduces to

$$T_=^-(1/2)\partial_- X_\mu.\partial_+ X^\mu$$

Further,

$$T_-^+ = (\partial L/\partial \partial_+ X^\mu).\partial_- X_\mu$$
$$+(\partial L/\partial \partial_+ \psi_-).\partial_- \psi_-$$
$$= (1/2)\partial_- X^\mu.\partial_- X_\mu + \psi_- \partial_- \psi_-$$

and likewise,

$$T_+^- = (1/2)\partial_+ X^\mu.\partial_+ X_\mu + \psi_+ \partial_+ \psi_+$$

Apart from the conservation law

$$\partial_+ T_+^+ + \partial_- T_+^- = 0, \partial_+ T_-^+ + \partial_- T_-^- = 0$$

we have another conservation law, namely the conservation of supersymmetry current which is the Noether current associated with the supersymmetry invariance of the Lagrangian density. This current can be derived as follows: Consider the local supersymmetry transformations

$$\delta X^\mu = k(x)^T \epsilon_0 \psi^\mu(x), \delta\psi^\mu(x) = c.\rho^\alpha k(x)\partial_\alpha X^\mu$$

where $k(x)$ is an infinitesimal Fermionic parameter field and c is a complex number. Then under this transformation, we obtain the following infinitesimal variations in the Bosonic and Fermionic components of the Lagrangian:

$$\delta((1/2)\partial_\alpha X^\mu.\partial^\alpha X_\mu) =$$

$$\partial_\alpha \delta X^\mu . \partial^\alpha X_\mu =$$
$$= k_{,\alpha}^T \epsilon_0 \psi^\mu \partial^\alpha X_\mu$$
$$+ k^T \epsilon_0 \psi_{,\alpha}^\mu \partial^\alpha X_\mu$$
$$\delta(-i\psi^{\mu T} \epsilon_0 \rho^\alpha \psi_{\mu,\alpha}) =$$
$$-i\delta\psi^{\mu T} \epsilon_0 \rho^\alpha \psi_{\mu,\alpha}$$
$$-i\psi^{\mu T} \epsilon_0 \rho^\alpha \delta\psi_{\mu,\alpha}$$
$$= -ick^T \rho^{\beta T} \epsilon_0 \rho^\alpha \psi_{\mu,\alpha} X_{,\beta}^\mu$$
$$-i\psi^{\mu T} \epsilon_0 \rho^\alpha c.\rho^\beta k X_{\mu,\alpha\beta}$$
$$-i\psi^{\mu T} \epsilon_0 \rho^\alpha c.\rho^\beta k_{,\beta} X_{\mu,\alpha}$$

After neglecting a perfect divergence (which does not contribute to the action integral), the change in total Lagrangian is then

$$\delta L =$$
$$k_{,\alpha}^T \epsilon_0 \psi^\mu \partial^\alpha X_\mu$$
$$+ k^T \epsilon_0 \psi_{,\alpha}^\mu \partial^\alpha X_\mu$$
$$-ick^T \rho^{\beta T} \epsilon_0 \rho^\alpha \psi_{\mu,\alpha} X_{,\beta}^\mu$$
$$+ic\psi_{,\alpha}^{\mu T} \epsilon_0 \rho^\alpha \rho^\beta k X_{\mu,\beta}$$
$$= k_{,\alpha}^T \epsilon_0 \psi^\mu \partial^\alpha X_\mu$$
$$+ k^T \epsilon_0 \psi_{,\alpha}^\mu \partial^\alpha X_\mu$$
$$-ick^T \rho^{\beta T} \epsilon_0 \rho^\alpha \psi_{\mu,\alpha} X_{,\beta}^\mu$$
$$+ic\psi_{,\alpha}^{\mu T} \epsilon_0 \rho^\alpha \rho^\beta k X_{\mu,\beta}$$
$$= k_{,\alpha}^T \epsilon_0 \psi^\mu \partial^\alpha X_\mu$$
$$+ k^T \epsilon_0 \psi_{,\alpha}^\mu \partial^\alpha X_\mu$$
$$-ick^T \rho^{\beta T} \epsilon_0 \rho^\alpha \psi_{\mu,\alpha} X_{,\beta}^\mu$$
$$+ic\psi_{,\alpha}^{\mu T} \epsilon_0 \rho^\alpha \rho^\beta k X_{\mu,\beta}$$
$$= k_{,\alpha}^T \epsilon_0 \psi^\mu \partial^\alpha X_\mu$$
$$+ k^T \epsilon_0 \psi_{,\alpha}^\mu \partial^\alpha X_\mu$$
$$+ick^T \epsilon_0 \rho^\beta \rho^\alpha \psi_{\mu,\alpha} X_{,\beta}^\mu$$
$$+ic\psi_{,\alpha}^{\mu T} \epsilon_0 \rho^\alpha \rho^\beta k X_{\mu,\beta}$$

since

$$\rho^{0T} \epsilon_0 = -\sigma_2.i.\sigma_2 = -iI_2, \epsilon_0 \rho^0 = i\sigma_2.\sigma_2 = iI_2$$

$$\rho^{1T}\epsilon_0 = i\sigma_1.i\sigma_2 = -i\sigma_3, \epsilon_0\rho^1 = i\sigma_2 i\sigma_1 = i\sigma_3$$

This is the same as saying that

$$\rho^{\alpha T}\epsilon_0 = -\epsilon_0\rho^\alpha$$

Now, using the anticommutativity of the Fermions k, ψ^μ and the antisymmetry of ϵ_0, we have that

$$\psi^{\mu T}_{,\alpha}\epsilon_0\rho^\alpha\rho^\beta kX_{\mu,\beta}$$

$$= k^T\rho^{\beta T}\rho^{\alpha T}\epsilon_0\psi_{\mu,\alpha}X^\mu_{,\beta}$$

$$= k^T\epsilon_0\rho^\beta\rho^\alpha\psi_{\mu,\alpha}X^\mu_{,\beta}$$

Thus, we get after neglecting a perfect divergence,

$$\delta L =$$

$$= k^T_{,\alpha}\epsilon_0\psi^\mu\partial^\alpha X_\mu$$

$$+ k^T\epsilon_0\psi^\mu_{,\alpha}\partial^\alpha X_\mu$$

$$+ 2ick^T\epsilon_0\rho^\beta\rho^\alpha\psi_{\mu,\alpha}X^\mu_{,\beta}$$

$$= k^T_{,\alpha}\epsilon_0\psi^\mu\partial^\alpha X_\mu$$

$$+ k^T\epsilon_0\psi^\mu_{,\alpha}\partial^\alpha X_\mu$$

$$- 2ick^T\epsilon_0\rho^\beta\rho^\alpha\psi_\mu X^\mu_{,\alpha\beta})$$

$$- 2ick^T_{,\alpha}\epsilon_0\rho^\beta\rho^\alpha\psi_\mu X^\mu_{,\beta}$$

$$= k^T_{,\alpha}\epsilon_0\psi^\mu\partial^\alpha X_\mu$$

$$+ k^T\epsilon_0\psi^\mu_{,\alpha}\partial^\alpha X_\mu$$

$$- ick^T\epsilon_0\{\rho^\beta, \rho^\alpha\}\psi_\mu X^\mu_{,\alpha\beta}$$

$$- 2ick^T_{,\alpha}\epsilon_0\rho^\beta\rho^\alpha\psi_\mu X^\mu_{,\beta}$$

Now,

$$\{\rho^\beta, \rho^\alpha\} = 2\eta^{\alpha\beta}I_2, \eta = diag[1, -1]$$

Thus,

$$\delta L =$$

$$= k^T_{,\alpha}\epsilon_0\psi^\mu\partial^\alpha X_\mu$$

$$+ k^T\epsilon_0\psi^\mu_{,\alpha}\partial^\alpha X_\mu$$

$$- 2ick^T\epsilon_0\psi_\mu\partial^\alpha\partial_\alpha X^\mu$$

$$- 2ick^T_{,\alpha}\epsilon_0\rho^\beta\rho^\alpha\psi_\mu X^\mu_{,\beta}$$

$$= k^T_{,\alpha}\epsilon_0\psi^\mu\partial^\alpha X_\mu$$

$$+k^T\epsilon_0\psi_{\mu,\alpha}\partial^\alpha X^\mu$$

$$+2ick^T_{,\alpha}\epsilon_0\psi_\mu\partial^\alpha X^\mu$$

$$+2ick^T\epsilon_0\psi_{\mu,\alpha}\partial^\alpha X^\mu$$

$$-2ick^T_{,\alpha}\epsilon_0\rho^\beta\rho^\alpha\psi_\mu X^\mu_{,\beta}$$

We now choose c so that $2ic+1=0$, ie,

$$c=-1/2i=i/2$$

Then, the above variation reduces to

$$\delta L=k^T_{,\alpha}\epsilon_0\rho^\beta\rho^\alpha\psi_\mu X^\mu_{,\beta}$$

In order to cancel this term to obtain local supersymmetry, as we shall see in a later chapter, additional terms involving a gaugino field must be added to this Lagrangian with an additional definition of its supersymmetry transformation rule.

[24] **Appendix**

This appendix presents some remarks about the Fermion Fock spaces which are required to develop the calculus of Fermionic superstrings, especially in determining the anticommutation relations between the Fourier series coefficients of a Fermionic superstring. We also present some general remarks about how gauge field supersymmetric Lagrangians are constructed using the theory of Chiral fields and their gauge transformations in supersymmetry theory. Some aspects of the matter field supersymmetric Lagrangians based on Chiral field theory are also presented. It also presents some general remarks about supersymmetric matter and gauge Lagrangians with a superpotential term whose parameters can be controlled to design a supersymmetric quantum unitary gate for use in a quantum computer. Another way of designing a quantum gate is presented based on breaking the supersymmetry of a gauge Lagrangian by introducing classical c-number currents interacting with the gauge field, just as the current density interacts with the four vector potential in electromagnetic field theory. The classical current sources are then designed so that the scattering matrix amplitudes computed on the basis of this broken gauge supersymmetric Lagrangian is as close as possible to a given unitary matrix. The scattering matrix computation is based on the Feynman path integral approach to quantum field theory.

A.1 **The Fermion Fock spaces**

[1] Fermion Fock space: Let V be an inner product space of finite dimension N and let $\wedge$ denote an antisymmetric tensor product on V. Consider the Grassmanian space

$$\Lambda V=\bigoplus_{k=0}^{N}\Lambda^k V,\Lambda^0 V=\mathbb{C}$$

For $u \in V$, define the linear operator $a(u)^*$ on ΛV by

$$a(u)^* w = u \wedge w, w \in \Lambda^n V, 0 \leq n \leq N$$

and the operator $a(u)$ on ΛV by

$$a(u)(v_1 \wedge ... \wedge v_n) = \sum_{k=1}^{n} (-1)^{k-1} < u, v_k > v_1 \wedge ... \hat{v}_k \wedge ... \wedge v_n$$

where by $\hat{v}$, we mean that v is omitted.

[a] Prove that $a(u)^*$ is the adjoint of $a(u)$

[b] Prove that

$$\{a(u), a(v)^*\} = < u, v >, a(u)^2 = 0, a(u)^{*2} = 0$$

[c] Define the Dirac Gamma operators by

$$\gamma(u) = a(u) + a(u)^*$$

Then show that

$$\{\gamma(u), \gamma(v)\} = 2 < u, v >$$

A.2. Compute $f(\Phi)$ explicitly as a fourth degree polynomial in θ where Φ is a general left Chiral field given by

$$\Phi(x, \theta) = \phi(x_+) + \theta_L^T \psi_L(x_+) + \theta_L^T \epsilon \theta_L F(x0$$

with

$$x_+^\mu = x^\mu + \theta_R^T \epsilon \gamma^\mu \theta_L$$

$$= x^\mu + \theta_L^T \epsilon \gamma^\mu \theta_R$$

A.3 Let $V(x, \theta)$ be an arbitrary superfield. Compute the components of the left Chiral field

$$W_L(x, \theta) = D_R^T \epsilon D_R D_L V(x, \theta)$$

assuming that V has been expressed in the Wess-Zumino gauge as

$$V(x, \theta) = \theta^T \epsilon \gamma^\mu \theta . V_\mu(x) + \theta^T \epsilon \theta \theta^T \gamma^5 \epsilon \lambda(x) + (\theta^T \epsilon \theta)^2 D(x)$$

Compute

$$Re[W_L^T \epsilon . W_L]_F$$

and show that it is Gauge invariant, supersymmetry invariant and also Lorentz invariant and is a linear combination of $f_{\mu\nu} f^{\mu\nu}, \lambda^T \gamma^5 \epsilon \lambda$ and D^2 where $f_{\mu\nu} = V_{\nu,\mu} - V_{\mu,\nu}$ and thus the action corresponding to this Lagrangian density is the supersymmetric generalization of the Maxwell Lagrangian density $f_{\mu\nu} f^{\mu\nu}$ with

V_μ being the Bosonic gauge field (ie the Maxwell four potential) and λ being the gaugino field, ie, the Fermionic superpartner of the Bosonic gauge field.

Note that a generalized gauge transformation of the gauge superfield V is given by

$$V \to exp(i\Omega)V.exp(-i\Omega^*)$$

where $\Omega = \Omega(\theta_L, x_+)$ is an arbitrary left Chiral superfield. If it is infinitesimal, then it results in an infinitesimal generalized gauge transformation

$$V \to V + i(\Omega - \Omega^*)$$

and we see that the left Chiral spinor superfield W_L is invariant under such a generalized gauge transformation since

$$(D_R^T\epsilon D_R)D_L(\Omega - \Omega^*) =$$

$$D_R^T\epsilon D_R D_L\Omega = 0$$

because firstly Ω is right Chiral and hence $D_L\Omega^* = 0$ and secondly, $D_R\Omega = 0$ and as seen earlier in these notes, $[D_R^T\epsilon D_R, D_{La}]$ is a linear combination of the commuting operators ∂_μ and D_{Rb}. Thus, We evaluate $[W_L^T\epsilon W_L]_F$ in the Wess-Zumino gauge: First,

$$D_L = \gamma^5\epsilon\partial_{\theta_L} - \gamma^\mu\theta_R\partial_\mu$$

$$\begin{aligned} D_R^T\epsilon D_R &= (\gamma^5\epsilon\partial_{\theta_R} - \gamma^\mu\theta_L\partial_\mu)^T\epsilon.(\gamma^5\epsilon\partial_{\theta_R} - \gamma^\mu\theta_L\partial_\mu) \\ &= \partial_{\theta_R}^T\epsilon.\partial_{\theta_R} + \theta_L^T\gamma^{\mu T}\epsilon.\gamma^\nu\theta_L\partial_\mu\partial_\nu \\ &\quad +\partial_{\theta_R}^T\gamma^5\epsilon\epsilon\gamma^\nu\theta_L\partial_\nu \\ &\quad -\theta_L^T\gamma^{\nu T}\epsilon.\gamma^5\epsilon\partial_{\theta_R}\partial_\nu \\ &= \partial_{\theta_R}^T\epsilon.\partial_{\theta_R} + \theta_L^T\epsilon\gamma^\mu.\gamma^\nu\theta_L\partial_\mu\partial_\nu \\ &\quad +\partial_{\theta_R}^T\gamma^\nu\theta_L\partial_\nu \\ &\quad -\theta_L^T\gamma^{\nu T}\partial_{\theta_R}\partial_\nu \\ &= \partial_{\theta_R}^T\epsilon.\partial_{\theta_R} + \theta_L^T\epsilon\theta_L.\Box \\ &\quad -2\theta_L^T\gamma^{\nu T}\partial_{\theta_R}\partial_\nu \end{aligned}$$

Now, we can write

$$\begin{aligned} V(x,\theta) &= \theta^T\epsilon\gamma^\mu\theta.V_\mu(x) + \theta^T\epsilon\theta\theta^T\gamma^5\epsilon\lambda(x) + (\theta^T\epsilon\theta)^2D(x) \\ &= 2\theta_L^T\epsilon\gamma^\mu\theta_R V_\mu + (\theta_L^T\epsilon\theta_L\theta_R^T + \theta_R^T\epsilon\theta_R\theta_L^T)\gamma^5\epsilon\lambda \\ &\quad +2(\theta_L^T\epsilon\theta_L)(\theta_R^T\epsilon\theta_R)D \end{aligned}$$

We first calculate the action of D_L on each of these terms and then the action of $D_R^T\epsilon D_R$ on each of the resulting terms.

$$D_L\theta_L^T\epsilon\gamma^\mu\theta_R V_\mu =$$

$$(\gamma^5\epsilon\partial_{\theta_L} - \gamma^\nu\theta_R\partial_\nu)\theta_L^T\epsilon\gamma^\mu\theta_R V_\mu =$$
$$\gamma^5\epsilon^2\gamma^\mu\theta_R V_\mu - \gamma^\nu\theta_R\theta_L^T\epsilon\gamma^\mu\theta_R V_{\mu,\nu}$$

Now,

$$\theta_R\theta_L^T = ((1-\gamma^5)/2)\theta.\theta^T((1+\gamma^5)/2)$$
$$= (1/4)\theta^T\epsilon\gamma^\mu\theta)((1-\gamma^5)/2)\gamma_\mu\epsilon((1+\gamma^5)/2)$$
$$= (1/2)\theta_L^T\epsilon\gamma^\mu\theta_R\gamma_\mu\epsilon P$$

where

$$P = (1+\gamma^5)/2$$

Thus,

$$D_L\theta_L^T\epsilon\gamma^\mu\theta_R V_\mu =$$
$$-\gamma^\mu\theta_R V_\mu - (1/2)\gamma^\nu\theta_L^T\epsilon\gamma^\rho\theta_R\gamma_\rho\epsilon^2\gamma^\mu\theta_R V_{\mu,\nu}$$
$$= -\gamma^\mu\theta_R V_\mu + (1/2)\theta_L^T\epsilon\gamma^\rho\theta_R\gamma^\nu\gamma_\rho\gamma^\mu\theta_R V_{\mu,\nu}$$

Then,

$$D_R^T\epsilon D_R D_L(\theta_L^T\epsilon\gamma^\mu\theta_R V_\mu)$$
$$= [\partial_{\theta_R}^T\epsilon.\partial_{\theta_R} + \theta_L^T\epsilon\theta_L.\Box$$
$$-2\theta_L^T\gamma^{\nu T}\partial_{\theta_R}\partial_\nu](-\gamma^\mu\theta_R V_\mu + (1/2)\theta_L^T\epsilon\gamma^\rho\theta_R\gamma^\nu\gamma_\rho\gamma^\mu\theta_R V_{\mu,\nu})$$

The term involving first order space-time derivatives of V_μ in this expression is given by

$$\partial_{\theta_R}^T\epsilon.\partial_{\theta_R}((1/2)\theta_L^T\epsilon\gamma^\rho\theta_R\gamma^\nu\gamma_\rho\gamma^\mu\theta_R V_{\mu,\nu})$$
$$+2\theta_L^T\gamma^{\nu T}\partial_{\theta_R}\partial_\nu(\gamma^\mu\theta_R V_\mu)$$

This expression is clearly linear in θ_L and does not contain θ_R. We first observe that

$$\partial_{\theta_R}^T\epsilon.\partial_{\theta_R}((1/2)\theta_L^T\epsilon\gamma^\rho\theta_R\gamma^\nu\gamma_\rho\gamma^\mu\theta_R V_{\mu,\nu})$$
$$= (1/2)\partial_{\theta_R}^T\epsilon.\partial_{\theta_R}(\gamma^\nu\gamma_\rho\gamma^\mu\theta_R\theta_R^T\epsilon\gamma^\rho\theta_L V_{\mu,\nu})$$
$$= \gamma^\nu\gamma_\rho\gamma^\mu\epsilon.\epsilon\gamma^\rho\theta_L V_{\mu,\nu}$$
$$= -\gamma^\nu\gamma_\rho\gamma^\mu\gamma^\rho\theta_L.V_{\mu,\nu}$$

where we have used the easily verifiable identity

$$(\partial_{\theta_R}^T\epsilon\partial_{\theta_R})(\theta_R\theta_R^T) =$$
$$2\epsilon(1-\gamma^5)/2$$

Next we observe that

$$2\theta_L^T\gamma^{\nu T}\partial_{\theta_R}\partial_\nu(\gamma^\mu\theta_R V_\mu)$$
$$= 2\gamma^\mu\gamma^\nu\theta_L.V_{\mu,\nu}$$

Combining these two calculations, we get that the term that containing only the first order space-time derivatives in $D_R^T\epsilon D_R D_L V$ is given by

$$-\gamma^\nu\gamma_\rho\gamma^\mu\gamma^\rho\theta_L.V_{\mu,\nu}+2\gamma^\mu\gamma^\nu\theta_L.V_{\mu,\nu}$$

Now,

$$\gamma^\nu\gamma_\rho\gamma^\mu\gamma^\rho=$$

$$\gamma^\nu\gamma_\rho(2\eta^{\mu\rho}-\gamma^\rho\gamma^\mu)$$

$$=2\gamma^\nu\gamma^\mu-\gamma^\nu\gamma_\rho\gamma^\rho\gamma^\mu$$

A simpler approach:We've seen that

$$D_L\theta_L^T\epsilon\gamma^\mu\theta_R V_\mu=$$

$$\gamma^5\epsilon^2\gamma^\mu\theta_R V_\mu-\gamma^\nu\theta_R\theta_L^T\epsilon\gamma^\mu\theta_R V_{\mu,\nu}$$

$$=\gamma^5\epsilon^2\gamma^\mu\theta_R V_\mu-\gamma^\nu\theta_R\theta_R^T\epsilon\gamma^\mu\theta_L V_{\mu,\nu}$$

and hence the term involving first order space-time derivatives in V_μ in $D_R^T\epsilon D_R D_L$ is given by

$$\partial_{\theta_R}^T\epsilon\partial_{\theta_R}(-\gamma^\nu\theta_R\theta_R^T\epsilon\gamma^\mu\theta_L V_{\mu,\nu})$$

$$+2\theta_L^T\gamma^{\nu T}\partial_{\theta_R}(-\gamma^\mu\theta_R V_\mu)$$

$$=[2\gamma^\nu\gamma^\mu\theta_L.V_{\mu,\nu}-2\gamma^\mu\gamma^\nu\theta_L]V_{\mu,\nu}$$

$$=-2[\gamma^\mu,\gamma^\nu]\theta_L.V_{\mu,\nu}=$$

$$[\gamma^\mu,\gamma^\nu]\theta_L f_{\mu\nu}(x)$$

where

$$f_{\mu\nu}=V_{\nu,\mu}-V_{\mu,\nu}$$

Some remarks: Any 4×4 skew-symmetric matrix X can be expanded as

$$X=c_\mu\gamma_\mu\epsilon+a\epsilon+b\gamma^5\epsilon$$

We get

$$Tr(X\epsilon\gamma^\mu)=c(\mu)Tr(\gamma_\mu\epsilon.\epsilon\gamma^\mu)$$

(No summation)

$$=-c(\mu)Tr(\gamma_\mu\gamma^\mu)=-2c(\mu)Tr(\sigma^\mu\sigma^\mu)$$

$$=-4c(\mu)$$

Exercises:

[1] The supersymmetric Abelian gauge field Lagrangian has the form

$$L = c_1 f_{\mu\nu} f^{\mu\nu} + c_2 \lambda^T \gamma^5 \epsilon \gamma^\mu \lambda_{,\mu} + c_3 D^2$$

where

$$f_{\mu\nu} = V_{\nu,\mu} - V_{\mu,\nu}$$

is the antisymmetric gauge field tensor and λ is its superpartner, ie, the gaugino field. D is an auxiliary field that is eliminated by the equations of motion. So effectively our gauge-gaugino action is

$$S[V_\mu, \lambda] = \int [f_{\mu\nu} f^{\mu\nu} + c_0 \lambda^T \gamma^5 \epsilon \gamma^\mu \lambda_{,\mu}] d^4x$$

Note that λ is a Majorana Fermionic field. This action is similar to what we have in quantum electrodynamics consisting of the Maxwell photon field built out of the Bosonic creation and annihilation operators of photons and the Dirac electron-positron field built out of the Fermionic creation and annihilation operators of the electrons and positrons. Expanding these fields in terms of these operators and making the Legendre transformation gives us the Hamiltonian in the form

$$H = \sum_k \omega(k) c(k)^* c(k) + \sum_k \eta(k)(a(k)^* a(k) + b(k)^* b(k)) - - - (1)$$

where $c(k)$ are photonic annihilation operators, $a(k)$ are the electron annihilation operators and $b(k)$ are the positron annihilation operators. Our evolution operator is then

$$U(t) = exp(-itH) = \Pi_k (exp(-it\omega(k)c(k)^* c(k)).exp(-it\eta(k)(a(k)^* a(k)).exp(-itb(k)$$

$$= \Pi_k exp(-it\omega(k)c(k)^* c(k))(\Pi_k((exp(-it\eta(k))-1)a(k)^* a(k)+1)(\Pi_k((exp(-it\eta(k))-1$$

Now, let $|n_1, n_2, ... >_{ph} = |\mathbf{n} >_{ph}$ denote the state in which there are n_k photons of the kth type ($n_k = 0, 1, 2, ...$)and let $|s_, s_2, ... >_e = |\mathbf{s} >_e$ denote the state in which there are s_k electrons of the kth type where $s_k = 0, 1$ and likewise $|r_1, r_2, ... >_p = |\mathbf{r}_p$ the state in which there are r_k positrons of the kth type where $r_k = 0, 1$. Then define

$$|\mathbf{n}, \mathbf{s}, \mathbf{r} >$$

to be the tensor product of these three states. We get

$$U(t)|\mathbf{n}, \mathbf{s}, \mathbf{r} >= exp(-it \sum_k (\omega(k) n_k + \eta(k)(s_k + r_k)))|\mathbf{n}, \mathbf{s}, \mathbf{r} >$$

So obviously, $U(t)$ is a diagonal operator in this basis. Now we introduce external c-number current (J^μ)and gauge potential fields A^μ which break the supersymmetry. The resulting Lagrangian density is given by

$$L = f_{\mu\nu} f^{\mu\nu} + c_0 \lambda^T \gamma^5 \epsilon \gamma^\mu \lambda_{,\mu}$$

$$+J^{\mu}V_{\mu}+A_{\mu}\lambda^{T}\gamma^{5}\epsilon\lambda_{,\mu}$$

We again expand V_{μ} as a linear combination of the $c(k)'s$ so that $f_{\mu\nu}$ is also a linear combination of the $c(k)'s$ and λ is a linear combination of the $a(k)'s$ and $b(k)'s$ and the total Hamiltonian corresponding to this action can be expressed in the form

$$H=H_0+\sum_k(J[k,t]c(k)+J[k,t]^*c(k)^*)+\sum_k(A_1[k,m,t]a(k)a(m)+A_2[k,m,t]a(k)^*a(m)+$$

$$A_3[k,m,t]b(k)b(m)+A_3[k,m,t]b(k)^*b(m)+cc)$$

where $J[k,t]$ are linear functionals of J^{μ} over the spatial variables at a given time t and $A_j[k,m,t]$ are linear functionals of A^{μ} over the spatial variables at a given time. we can now apply perturbation theory or equivalently the Dyson series in the interaction picture to design our quantum gate.

[2] Consider the Chiral matter Lagrangian defined by

$$L=[\Phi^*\Phi]_D$$

where Φ is a left Chiral field defined by

$$\Phi(x)=\phi(x_+)+\theta_L^T\epsilon\psi_L(x_+)+\theta_L^T\epsilon\theta_LF(x)$$

Construct L explicitly and derive its form as a function of the scalar and left Dirac fields ϕ,ψ_L after eliminating the auxiliary field F. Then quantize this theory using perturbation theory. We have above evaluated

$$L=[\Phi^*\Phi]_D$$

First we computed

$$\Phi^*=\phi^*(x_-)+\theta_R^T\gamma^5\epsilon\gamma^0\epsilon\psi^*(x_-)+\theta_R^T\gamma^5\epsilon\gamma^0\epsilon\gamma^5\epsilon\gamma^0\theta_RF^*(x)$$

$$=\phi^*(x_-)+\theta_R^T\gamma^0\psi^*(x_)+\theta_R^T\epsilon\theta_RF^*(x)$$

Then,

$$[\phi^*(x_-)\phi(x_+)]_D=T_1+T_2+T_3$$

where

$$T_1=c_1\partial_{\mu}\phi^*(x).\partial^{\mu}\phi(x)$$
$$T_2=c_1\phi^*(x)\Box\phi(x)$$
$$T_3=c_1\phi(x)\Box\phi^*(x)$$

Further, we evaluated

$$[\theta_R^T\gamma^0\psi^*(x_-)\theta_L^T\epsilon\psi(x_+)]_{\theta^4}$$
$$=-2c_1(\gamma^{\mu}\psi_{,\mu}(x))^*\gamma^0\psi(x)$$

where we have used the fact that $\gamma_{\mu}\gamma^0$ is Hermitian and hence its transpose $\gamma^0\gamma_{\mu}^T$ is also Hermitian. Note that the conjugate of this quantity is given by

$$-2c_1\psi(x)^*\gamma^0\gamma^{\mu}\psi_{,\mu}(x)=-2c_1\bar{\psi}(x)\gamma^{\mu}\psi_{,\mu}(x)$$

$$= -2c_1\psi(x)^*\alpha^\mu\psi_{,\mu}(x)$$

Finally, we easily see that

$$[\phi^*(x_-)\theta_L^T\epsilon\psi(x_+)]_{\theta^4} = 0$$

Thus, assuming that ϕ is a real field, we get for the Lagrangian with neglect of perfect four divergence terms (which do not contribute anything to the action),

$$[\Phi^*\Phi]_D = c_1\partial_\mu\phi\partial^\mu\phi + c_2\psi(x)^*\alpha^\mu\psi_{,\mu}$$

Note that ψ is a left handed Majorana Fermion and hence

$$\psi(x)^*\alpha^\mu = \psi(x)^*\gamma^0\gamma^\mu = \psi(x)^T\gamma^5\epsilon$$

and hence the Fermionic contribution to this Lagrangian may also be expressed as

$$c_1\psi(x)^T\gamma^5\epsilon.\gamma^\mu\psi_{,\mu}$$

The last term in $[\Phi^*\Phi]_{\theta^4}$ is

$$F^*F = F^2$$

assuming F to be real. So far the Lagrangian has come out to be just the sum of the Lagrangians of the scalar Klein-Gordon field and that of the Dirac field, decoupled from each other and both with zero mass. There is no new supersymmetric effect. To get supersymmetric effects, we introduce a superpotential and add the supersymmetric Lagrangian $[f(\Phi)]_F$ to it. We get

$$\Phi(x) = \phi(x_+) + \theta_L^T\epsilon\psi_L(x_+) + \theta_L^T\epsilon\theta_L F(x)$$

$$= \phi(x)+\phi_{,\mu}(x)\theta_R^T\epsilon\gamma^\mu\theta_L+c_1\Box\phi(x)(\theta^T\epsilon\theta)^2+\theta_L^T\epsilon\psi(x)+\theta_R^T\epsilon\gamma^\mu\theta_L\theta_L^T\epsilon.\psi_{,\mu}(x)+\theta_L^T\epsilon\theta_L F(x)$$

$$[f(\Phi)]_F = f'(\phi)F + (1/2)f''(\phi)(\theta_L^T\epsilon\psi)^2$$
$$= f'(\phi)F + (1/2)f''(\phi)\psi^T\epsilon\theta_L\theta_L^T\epsilon\psi$$
$$= f'(\phi)F + c_2f''(\phi)\psi^T\gamma^5\epsilon\psi$$

Note that since ψ is left handed,

$$\gamma^5\psi = \psi$$

so since ϵ and γ^5 commute,

$$\psi^T\epsilon\psi = \psi^T\gamma^5\epsilon\psi$$

Since ψ is Majorana Fermion, we can also write

$$\psi^T\gamma^5\epsilon\psi = \psi^*\gamma^0\psi$$

and hence this term is a mass term in the Lagrangian with mass proportional to $f''(\phi)$. Finally, we can include a supersymmetric gauge-matter interaction term in the Lagrangian given by

$$L_{gm} = [\Phi^*\Gamma.\Phi]_D$$

where

$$\Gamma = \Gamma(x, \theta)$$

is the gauge superfield that transforms under a gauge transformation as

$$\Gamma \to exp(i\Omega^*)\Gamma.exp(-i\Omega)$$

with the matter superfield Φ transforming under the gauge transformation as

$$\Phi \to exp(i\Omega)\Phi$$

and hence,

$$\Phi^* \to \Phi^*.exp(-i\Omega^*)$$

where Ω is an arbitary left Chiral superfield. Thus, $\Phi^*\Gamma\Phi$ and hence $[\Phi^*\Gamma\Phi]_D$ is invariant under this generalized gauge transformation. We evaluate this choosing a Wess-Zumino gauge for Γ: (Note that Γ can always be brought to the Wess-Zumino gauge by an appropriate choice of the generalized gauge transformation Chiral superfield Ω and once it is in the Wess-Zumino gauge, it will always remain so under a restricted Gauge transformation, ie, with $\Omega = \Lambda(x_+)$. In the Wess-Zumino gauge,

$$\Gamma = \theta^T \epsilon\gamma^\mu\theta.V_\mu + \theta^T \epsilon\theta.\theta^T\gamma^5\epsilon\lambda + b(\theta\epsilon\theta)^2\Box D$$

Then recalling that

$$\Phi(x) = \phi(x_+) + \theta_L^T\epsilon\psi_L(x_+) + \theta_L^T\epsilon\theta_L F(x)$$

$$\Phi^* = \phi^*(x_-) + \theta_R^T\gamma^0\psi^*(x_-) + \theta_R^T\epsilon\theta_R F^*(x)$$

we get that for real ϕ,

$$[\phi(x_-)\Gamma\phi(x_+)]_{\theta^4} =$$

$$\phi(x)[\theta^T\epsilon\gamma^\mu\theta V_\mu\theta_R^T\epsilon\gamma^\nu\theta_L]_4\phi_{,\nu}(x) + b\phi(x)^2 D$$

$$-\phi_{,\nu}(x)[\theta^T\epsilon\gamma^\mu\theta V_\mu\theta_R^T\epsilon\gamma^\nu\theta_L]_4\phi(x)$$

$$= 4\phi^2 D$$

apart from a constant multiplicative factor.This is not very interesting. So we assume ϕ to be a complex scalar field in order to obtain interaction terms between the scalar field current and the vector potential V_μ. The term $\phi^2 D$ gives an interaction between the scalar field and the auxiliary field. In case that ϕ is complex, we would have to replace this term with

$$[\phi^*(x_-)\Gamma\phi(x_+)]_4 =$$

$$-\phi^*_{,\nu}(x)[\theta_R^T\epsilon\gamma^\nu\theta_L\theta_R^T\epsilon\gamma^\mu V_\mu]_4\phi(x) + b\phi(x)^*\phi(x)D$$

$$+\phi^*(x)[\theta_R^T\epsilon\gamma^\mu\theta_L.V_\mu\theta_R^T\epsilon\gamma^\nu\theta_L]_4\phi_{,\nu}(x)$$

$$= [\phi^*\phi_{,\mu} - \phi\phi^*_{,\mu}]V^\mu + b\phi^*\phi.D$$

Likewise,

$$[\theta_R^T\gamma^0\psi^*(x_-)\Gamma.\theta_L^T\epsilon\psi(x_+)]_4$$
$$= [\psi^*(x)^T\gamma^0\theta_R\theta_R^T\epsilon\gamma^\mu\theta_L.V_\mu\theta_L^T\epsilon\psi(x)]_4$$
$$= \psi^*(x)\gamma^0\gamma^\mu\psi(x)V_\mu$$

except for a multiplicative constant. This term represents the interaction between the Dirac current and the electromagnetic field appearing in quantum electrodynamics. Further,

$$[\phi^*(x_-)\Gamma\theta_L^T\epsilon\psi(x_+)]_4 =$$
$$= \phi(x)^*[\theta^T\epsilon\theta\theta^T\gamma^5\epsilon\lambda\theta_L^T\epsilon]_4\psi(x)$$
$$= \phi(x)^*[\theta_R^T\epsilon\theta_R\theta_L^T\gamma^5\epsilon\lambda\theta_L^T\epsilon]_4\psi(x)$$
$$= \phi(x)^*[\theta_R^T\epsilon\theta_R\theta_L^T\epsilon\lambda\theta_L^T\epsilon]_4\psi(x)$$
$$= -\phi(x)^*[\theta_R^T\epsilon\theta_R\psi(x)^T\epsilon\theta_L\theta_L^T]_4\epsilon\lambda$$
$$= \phi(x)^*\psi(x)^T\epsilon\lambda(x)$$

apart from a multiplicative constant. Note that since ψ is lefthanded, we can assume without loss of generality that λ is also leftthanded, ie,

$$\lambda = ((1+\gamma^5)/2)\lambda$$

This term is clearly a supersymmetric correction term to the Lagrangian comprising of the scalar field ϕ, the Dirac fields ψ, λ, and the electromagnetic field. The total supersymmetric Lagrangian for the matter and gauge fields taking interactions into account therefore has the form after eliminating the auxiliary fields F, D by using their field equations is given by (Note that before eliminating the auxiliary fields, the Lagrangian is supersymmetric but after their elimination, it is no longer supersymmetric)

$$L = (1/2)\partial_\mu\phi.\partial^\mu\phi + c_0 Im(\phi^*\phi_{,\mu})V^\mu + c_1\psi^T\gamma^5\epsilon\gamma^\mu\psi_{,\mu}$$
$$c_2 f_{\mu\nu}f^{\mu\nu} + c_3\lambda^T\gamma^5\epsilon\gamma^\mu\lambda_{,\mu}$$
$$+c_4 f''(\phi)\psi^T\gamma^5\epsilon\psi + c_5\psi(x)^T\epsilon\gamma^\mu\psi(x)V_\mu$$
$$+c_6 Re(\phi(x)^*\psi(x)^T\epsilon\lambda(x))$$

Note that if the auxiliary terms were included here, their total contribution would be

$$c_7F^*F + c_8|\phi|^2D + c_9D^2 + 2c_{10}Re(f'(\phi)F)$$

where the term F^F comes from $[\Phi^*\Phi]_D$, the term $|\phi|^2D$ comes from $[\Phi^*\Gamma.\Phi]_4$, the term D^2 comes from the pure gauge term $Re[W_L^T\epsilon W_L]_D$ with $W_L = D_R^T\epsilon D_R D_L V$ and finally, the term $f'(\phi)F$ comes from $[f(\Phi)]_F$. The field equations for D, F are therefore given by

$$2c_7F + c_{10}f'(\phi)^* = 0, 2c_9D + c_8|\phi|^2 = 0$$

Substituting for these auxiliary fields into the above supersymmetric Lagrangian gives us the effective Lagrangian for $(\phi, \psi, V_\mu, \lambda)$ as

$$L = \partial_\mu \phi^*.\partial^\mu \phi + c_0 Im(\phi^* \phi_{,\mu})V^\mu + c_1 \psi^T \gamma^5 \epsilon \gamma^\mu \psi_{,\mu}$$

$$c_2 f_{\mu\nu} f^{\mu\nu} + c_3 \lambda^T \gamma^5 \epsilon \gamma^\mu \lambda_{,\mu}$$

$$+c_4 f''(\phi)\psi^T \gamma^5 \epsilon \psi + c_5 \psi(x)^T \epsilon \gamma^\mu \psi(x) V_\mu$$

$$+c_6 Re(\phi(x)^* \psi(x)^T \epsilon \lambda(x))$$

$$+c_7' |f'(\phi)|^2 + c_8' (\phi(x)^* \phi(x))^2$$

and that completes the story of our supersymmetric Lagrangian with the auxiliary fields eliminated. Note that ψ and λ are left handed Fermionic fields. Taking in addition, external currents into consideration via an interaction term $J^\mu V_\mu$ gives us control over the other quantum fields which can then be used to design a gate.

The field equations obtained by setting the variational derivatives of L w.r.t $\phi, \psi, V_\mu, \lambda$ are given by

$$-\Box\phi + c_4 f'''(\phi)\psi^T \gamma^5 \epsilon \psi + c_6 \psi^T \epsilon \gamma^\mu \lambda_{,\mu}$$

$$+c_7' f''(\phi)^* f'(\phi) + 2c_8' \phi^* \phi^2 = 0$$

$$2c_1 \gamma^5 \epsilon \gamma^\mu \psi_{,\mu} + 2c_4 f''(\phi)\gamma^5 \epsilon \psi$$

$$+2c_5 V_\mu \gamma^5 \epsilon \gamma^\mu \psi + c_6 \phi^*.\gamma^5 \epsilon \lambda = 0$$

or equivalently,

$$c_1 \gamma^\mu \psi_{,\mu} + c_4 f''(\phi)\psi + c_5 V_\mu \gamma^\mu \psi + c_6 \phi^*.\lambda = 0,$$

$$2c_3 \gamma^5 \epsilon \gamma^\mu \lambda_{,\mu} - c_6 \phi^* \gamma^5 \epsilon \psi = 0$$

or equivalently,

$$c_3 \gamma^\mu \lambda_{,\mu} - c_6 \phi^* \psi = 0$$

Remark:The above Lagrangian can also be written as noting that ψ is left handed,

$$L = (1/2)\partial_\mu \phi.\partial^\mu \phi + c_0 Im(\phi^* \phi_{,\mu})V^\mu + c_1 \psi^T \gamma^5 \epsilon \gamma^\mu \psi_{,\mu}$$

$$c_2 f_{\mu\nu} f^{\mu\nu} + c_3 \lambda^T \gamma^5 \epsilon \gamma^\mu \lambda_{,\mu}$$

$$+c_4 f''(\phi)\psi^T \gamma^5 \epsilon \psi + c_5 \psi(x)^T \gamma^5 \epsilon \gamma^\mu \psi(x) V_\mu$$

$$+c_6 Re(\phi(x)^* \psi(x)^T \gamma^5 \epsilon \lambda(x))$$

The above supersymmetric Lagrangian after eliminating the auxiliary fields is denoted by $L(\phi, \psi, V_\mu, \lambda)$. It depends on a control potential f. We parametrize f by a set of parameters $\theta = (\theta_1, ..., \theta_p)$, so the Lagrangian can be written as

$$L(\chi|\theta), \chi = (\phi, \psi, V_\mu, \lambda)$$

The path integral for this action can be used to compute the S-matrix element between two states as follows. Let $|0>$ be the vacuum state and then an arbitrary initial state can be expressed as

$$|i>= G_i(\chi(t_i))|0>$$

where $G_i(\chi(t_i))$ is some functional of $(\chi(x,t_i) : x \in \mathbb{R}^3)$ and t_i is the initial time. Likewise the final state can be expressed as

$$|f>= G_f(\chi(t_f))|0>$$

where

$$\chi(t_f) = (\chi(x,t_f) : x \in \mathbb{R}^3)$$

and hence the transition probability amplitude between the states $|i>$ and $|f>$ can be expressed as a path integral

$$S(f,i|\theta) =< f|S(\theta)|i>=$$

$$\int exp(i\int_{\mathbb{R}^3\times[t_i,t_f]} L(\chi(x,t)|\theta)d^3xdt)G_f(\chi(t_f))^*G_i(\chi(t_i))\Pi_{x\in\mathbb{R}^3,t_i\leq t\leq t_f}d\chi(x,t),$$

and to design a gate having matrix given matrix elements $U(f,i)$ for a class of pairs of initial and final states $(i,f) \in C$, we must choose θ so that

$$E(\theta) = \sum_{(i,f)\in C} |S(f,i|\theta) - U(f,i)|^2$$

is a minimum. Thus, without breaking supersymmetry by just adjusting the superpotential parameters, we can design an optimal unitary gate.

Supersymmetric non-Abelian gauge field Lagrangians. Let $V^A(x,\theta)$ be the gauge superfields in the Wess-Zumino gauge with the index A running over the non-Abelian gauge group generator indices. Thus, if t_A are the Hermitian Lie algebra generators of the gauge group, we define

$$\Gamma(x,\theta) = exp(\sum_A t^A V_A(x,\theta)) = exp(t.V) = 1 + t_A V^A + t_A t_B V^A V^B$$

since V^A contains only of second, third and fourth degrees in the Fermionic variables θ. Its transformation law under a generalized gauge transformation is defined by

$$\Gamma \to exp(i\Omega^*)\Gamma.exp(-i\Omega)$$

where Ω is an arbitrary left Chiral superfield. We also note that under such a generalized gauge transformation, Φ transforms as

$$\Phi \to exp(i\Omega)\Phi$$

so

$$\Phi^* \to \Phi^*.exp(-i\Omega^*)$$

and hence

$$\Phi^*\Gamma.\Phi \rightarrow \Phi^*.exp(-i\Omega^*).exp(i\Omega^*).\Gamma.exp(-i\Omega).exp(i\Omega)\Phi$$

$$= \Phi^*\Gamma.\Phi$$

proving gauge invariance of the Lagrangian. We define the generalized gauge field spinor by

$$W_L(x,\theta) = D_R^T \epsilon D_R exp(-t.V) D_L exp(t.V)$$

Under the above generalized gauge transformation, this transforms as

$$W_L \rightarrow D_R^T \epsilon D_R exp(i\Omega).exp(-t.V).exp(-i\Omega^*)(D_L.exp(i\Omega^*)exp(t.V).exp(-i\Omega))$$

$$= exp(i\Omega) D_R^T \epsilon D_R exp(-t.V) D_L.(exp(t.V)exp(-i\Omega))$$

and since

$$D_R^T \epsilon D_R D_L exp(-i\Omega) = [D_R^T \epsilon D_R, D_L].exp(-i\Omega) = 0$$

because Ω is left Chiral and hence $D_R exp(-i\Omega) = 0$ and further, $[D_R^T \epsilon D_R, D_L]$ is a linear combination of the $D'_R s$ with coefficients that are Bosonic differential operators, it follows that under the generalized gauge transformation given above,

$$W_L \rightarrow exp(i\Omega).W_L.exp(-i\Omega)$$

(Note that both sides here are left Chiral superfields).

Properties and computation of the supercurrent.

First we discuss a very general aspect of Noether's current conservation theorem applicable to any Lagrangian density for a set of fields having a symmetry. Let $\mathcal{L}(\phi_l, \phi_{l,\mu})$ be a Lagrangian density for the fields $\phi_l, l = 1, 2, ..., N$ depending only on their values and their first order space-time partial derivatives. Assume that there exists a group of transformations of the fields such that if an infinitesimal transformation of these fields is given by

$$\epsilon\delta\phi_l(x) = \epsilon.F_l(\phi_l(x), \phi_{l,\mu}(x))$$

where ϵ is a small parameter, then the change in the Lagrangian density under this group has the form

$$\delta\mathcal{L} = \frac{\partial\mathcal{L}}{\partial\phi_l}\delta\phi_l + \frac{\partial\mathcal{L}}{\partial\phi_{l,\mu}}\delta\phi_{l,\mu}$$

$$= \partial_\nu F^\nu(\phi_l, \phi_{l,\mu})$$

ie, the functions F^ν are functions of only the fields and their first order partial derivatives.

Chapter 3

Interaction Between Light and Matter in a Cavity of Arbitrary Shape

[1] Some Models for the interaction between light and matter based on non-Abelian gauge theories for the matter component.

Let $B^a_\mu(x)$ be non-Abelian gauge fields and $A_\mu(x)$ be the photon field. The matter wave function $\psi(x)$ satisfies the matter wave equation

$$[i\gamma^\mu\nabla_\mu - m]\psi(x) = 0$$

where

$$\nabla_\mu = \partial_\mu - ieA_\mu - ieB^a_\mu\tau_a$$

with τ_a being the Hermitian generators of the gauge group $G \subset U(n)$. In turn, the matter currents associated with the photon field and gauge field are

$$J^\mu(x) = -e\psi(x)^*\gamma^0\gamma^\mu\psi(x),$$

$$J^\mu_a(x) = -e\psi(x)^*(\gamma^0\gamma^\mu \otimes \tau_a)\psi(x)$$

and the matter field equations are

$$\partial_\nu F^{\mu\nu} = J^\mu,$$

$$(D_\nu F^{\mu\nu})_a = J^\mu_a$$

where

$$D_\nu F^{\mu\nu}_a = \partial_\nu F^{\mu\nu}_a + eC(abc)B^b_\nu F^{\mu\nu c}$$

with $C(abc)$ denoting the structure constants of the gauge group and

$$F^{\mu\nu a} = B^a_{\nu,\mu} - B^a_{\mu,\nu} + eC(abc)B^b_\mu B^c_\nu$$

We write

$$B^a_\mu = B^{0a}_\mu + \delta B^a_\mu$$

where B^{0a}_μ is the classical background Yang-Mills gauge field assumed to be classical and δB^a_μ is the small quantum fluctuation in this field. We then express the Lagrangian density of the Yang-Mills gauge field as

$$F^a_{\mu\nu}F^{\mu\nu a} = (F^{0a}_{\mu\nu} + \delta F^a_{\mu\nu}).(F^{0\mu\nu a} + \delta F^{\mu\nu a})$$

where

$$F^{0a}_{\mu\nu} = B^{0a}_{\nu,\mu} - B^{0a}_{\mu,\nu} + eC(abc)B^{0b}_\mu B^{0c}_\nu$$

is the classical component, ie, background Yang-Mills field tensor and

$$\delta F^a_{\mu\nu} = \delta B^a_{\nu,\mu} - \delta B^a_{\mu,\nu} + eC(abc)(B^{0b}_\mu \delta B^c_\nu + \delta B^b_\mu.B^{0c}_\nu) + eC(abc)\delta B^b_\mu \delta B^c_\nu$$

is the purely quantum component of the Yang-Mills field tensor. The linear-quadratic part of the Lagrangian density of the Yang-Mills gauge field is given by

$$L_{quadYM} = (\delta B^a_{\nu,\mu} - \delta B^a_{\mu,\nu}).(\delta B^{\nu,\mu a} - \delta B^{\mu,\nu a})$$
$$+2F^{0a}_{\mu\nu}(\delta B^a_{\nu,\mu} - \delta B^a_{\mu,\nu})$$
$$+2C(abc)F^{0a}_{\mu\nu}\delta B^b_\mu \delta B^c_\nu$$

We can easily compute the propagator of this quantum component of the Lagrangian density. In fact, if we denote the fields $\delta B^a_\mu(x)$ by $\phi_k(x), k = 1, 2, ..., N$, then this quadratic part of the Lagrangian density has the general form

$$L_q = \int K_{ij}(x,y)\phi_i(x)\phi_j(y)d^4xd^4y + \int M_i(x)\phi_i(x)d^4x$$

and we can compute easily the propagator of ϕ as a Gaussian field:

$$D_\phi(x,y) = \int exp(iS_q)D\phi =$$

$$(det(iK/2))^{1/2}exp((i/4)\int M_i(x)K_{ij}(x,y)M_j(y)d^4xd^4y)$$

We can now analyze the effects of cubic and four degree terms on the gauge field propagator using perturbation theory for path integrals according to the method laid out by Richard Feynman: We write the cubic and quadratic perturbations to the above Lagrangian density as

$$\delta L_q = \int P_{ijk}(x,y,z)\phi_i(x)\phi_j(y)\phi_k(z)d^4xd^4yd^4z$$
$$+\int Q_{ijkm}(x,y,z,v)\phi_i(x)\phi_j(y)\phi_k(z)\phi_m(v)^4xd^4yd^4zd^4v$$

Note that these cubic and fourth degree terms represent the terms

$$eC(abc)(\delta B^a_{\nu,\mu} - \delta B^a_{\mu,\nu})(\delta B^{\mu b}\delta B^{\nu c})$$

$$+e^2C(abc)C(apq)\delta B_\mu^b \delta B_\nu^c \delta B^{\mu p}\delta B^{\nu q}$$

Suppose that in principle, we have derived an expression for this gauge propagator. We can then as the question, what are the moments of the gauge field on the retinal screen for an initial state $\rho(0)$ of the matter plus gauge field ? As before, ψ denotes the matter field and ϕ the gauge field. The total Lagrangian of the matter plus gauge field has the form

$$L = \int M(x)^T\phi(x)d^4x + \int \phi(x)^T K(x,y)\phi(y)d^4xd^4y$$

$$+\int P(x,y,z)^T(\phi(x)\otimes\phi(y)\otimes\phi(z))d^4xd^4yd^4z$$

$$+\int Q(x,y,z,v)^T(\phi(x)\otimes\phi(y)\otimes\phi(z)\otimes\phi(v))d^4xd^4yd^4zd^4v$$

$$+\int \psi(x)^*R(x,y)\psi(y)d^4xd^4y + \int \psi(x)^*S_k(x)\psi(x)\phi_k(x)d^4x$$

The path integral with time ranging over $[0,T]$ results in an evolution kernel $U_T(\phi_f,\psi_f|\phi_i,\psi_i)$ where ϕ_i,ψ_i are gauge and matter fields over space at time $t=0$ while ϕ_f,ψ_f are gauge and matter fields over space at time $t=T$. By space, we mean the spatial volume region of the CDRA. Now suppose that at time $t=0$, the state of the matter and gauge field is represented by the density matrix kernel $\rho_0(\phi_1,\psi_1;\phi_2,\psi_2)$. Then the state of these fields at time $t=T$ will be given by

$$\rho_T(\phi_1,\psi_1;\phi_2,\psi_2) =$$

$$\int U_T(\phi_1,\psi_1|\phi_3,\psi_3)\rho_0(\phi_3,\psi_3;\phi_4,\psi_4)\bar{U}_T(\phi_2,\psi_2;\phi_4,\psi_4)D\phi_3 D\psi_3 D\phi_4 D\psi_4$$

Once we know this state at time T, we are in a position to calculate the statistical moments of the gauge field on the screen at time T.

[2] **Entanglement of the Fermionic modes caused by interaction with the photon radiation field**

When the electron-positron field interacts with the Bosonic radiation field, the Dirac current density acquires extra terms involving coupling between the Fermionic and Bosonic components. We analyze this interaction in what follows. The Dirac equation in the presence of the radiation field is given by

$$[\gamma^\mu(i\partial_\mu + eA_\mu) - m]\psi = 0$$

or equivalently,

$$[i\gamma^\mu\partial_\mu - m]\psi = -e\gamma^\mu A_\mu\psi$$

If $\psi^{(0)}$ denotes the free Dirac field, then we can write down the approximate solution to the above equation based on first order perturbation theory as

$$\psi(x) = \psi^{(0}(x) - e\int S_e(x-y)\gamma^\mu A_\mu(y)\psi^{(0)}(y)d^4y = \psi^{(0)}(x) + \psi^{(1)}(x)$$

say, where $S_e(x-y)$ is the electron propagator defined by

$$S_e(p) = \int S_e(x)exp(-ip.x)d^4x = i\gamma^\mu p_{mu} - m = i\gamma.p - m$$

Now writing the cavity constrained free Dirac field as

$$\psi^{(0)}(x) = \sum_k [b(k)\chi_k(x) + c(k)^*\eta_k(x)]$$

and the free cavity constrained electromagnetic four potential as

$$A_\mu(x) = \sum_k [a(k)\theta_k(x) + a(k)^*\theta_k(x)^*]$$

where $a(k), a(k)^*$ are the photon annihilation and creation operators while $b(k), b(k)^*$ are the electron annihilation and creation operators and $c(k), c(k)^*$ are the positron annihilation and creation operators, we get for the approximate value of the Dirac current operator,

$$J^\mu = (\psi^{(0)} + \psi^{(1)})^*\alpha^\mu(\psi^{(0)} + \psi^{(1)}) =$$

$$J^{\mu(0)} + \delta J^\mu$$

where

$$J^{\mu(0)}(x) = \psi^{(0)*}\alpha^\mu\psi^{(0)},$$

$$\delta J^\mu = \psi^{(0)*}\alpha^\mu\psi^{(1)} + \psi^{(1)*}\alpha^\mu\psi^{(0)}$$

Now,

$$J^{\mu(0)} = \sum_{km} (b(k)\chi_k(x) + c(k)^*\eta_k(x))^*\alpha^\mu(b(m)\chi_m(x) + c(m)^*\eta_m(x))$$

$$= \sum_{k,m} [b(k)^*b(m)\chi_k(x)^*\alpha^\mu\chi_m(x) + c(k)c(m)^*\eta_k(x)^*\alpha^\mu\eta_m(x)$$

$$+ b(k)^*c(m)^*\chi_k(x)^*\alpha^\mu\eta_m(x) + c(k)c(m)^*\eta_k(x)^*\alpha^\mu\eta_m(x)]$$

is the free Dirac current ie, in the absence of interactions with the photon field. We have already indicated how to compute the far field radiation pattern produced by this field and how to evaluate the moments of this field. Specifically, if $G(x-y)$ denotes the causal Green's function for the wave operator, then the electromangetic four potential produced by the Dirac current is given by

$$A_\mu(x) = \int G(x-y)J^\mu(y)d^4y = \int G(x-y)J^{\mu(0)}(y)d^4y + \int G(x-y)\delta J^\mu(y)d^4y = A_\mu^{(0)}(x) + \delta A_\mu(x)$$

Remark: We can consider a Fermionic coherent state rather than a Fermionic number state. Such a state is parametrized by a Grassmannian vector variable $\gamma = (\gamma_b(k), \gamma_c(k))_k$ and is denoted by $|\phi(\gamma)>$. The action of the electron and positron annihilation operators on this state is

$$b(k)|\phi(\gamma)> = \gamma_b(k)|\phi(\gamma)>,$$

$$c(k)|\phi(\gamma)> = \gamma_c(k)|\phi(\gamma)>$$

In order that the CAR

$$[b(k),b(m)]_+ = [c[k],c(m)]_+ = [b(k),c(m)]_+ = 0$$

hold good, we require that the Grassmannian parameters satisfy the anticommutation rules

$$\gamma_b(k)\gamma_b(m) + \gamma_b(m)\gamma_b(k) = 0,$$

$$\gamma_c(k)\gamma_c(m) + \gamma_c(m)\gamma_c(k) = 0,$$

$$\gamma_b(k)\gamma_c(m) + \gamma_c(m)\gamma_b(k) = 0,$$

Further, by our analogy with Bosonic coherent states, we impose the requirement that

$$[\partial/\partial\gamma_r(k),\gamma_s(m)]_+ = \delta(r,s)\delta(k,m), r,s = b,c$$

and that

$$b(k)^*|\phi(\gamma)>= (\partial/\partial\gamma_b(k))|\phi(\gamma)>,$$

$$c(k)^*|\phi(\gamma)>= (\partial/\partial\gamma_c(k))|\phi(\gamma)>$$

We then get

$$<\phi(\gamma)|b(k)^*b(m)|\phi(\gamma)>=<\phi(\gamma)|b(k)^*\gamma_b(m)|\phi(\gamma)>$$

$$=<b(k)\phi(\gamma)|\gamma_b(m)|\phi(\gamma)>=\gamma_b(k)^*\gamma_b(m)$$

on the one hand, while on the other,

$$<\phi(\gamma)|b(m)b(k)^*|\phi(\gamma)>=$$

$$<\phi(\gamma)|b(m)(\partial/\partial\gamma_b(k))|\phi(\gamma)>=$$

$$<\phi(\gamma)|(\partial/\partial\gamma_b(k))b(m)|\phi(\gamma)>$$

$$=<\phi(\gamma)|(\partial/\partial\gamma_b(k))\gamma_b(m)|\phi(\gamma)>$$

$$=\delta(k,m)-<\phi(\gamma)|\gamma_b(m)(\partial/\partial\gamma_b(k))|\phi(\gamma)>$$

$$=\delta(k,m)-\gamma_b(m)<\phi(\gamma)|b(k)^*|\phi(\gamma)>$$

$$=\delta(k,m)-\gamma_b(m)<b(k)\phi(\gamma)|\phi(\gamma)>=$$

$$\delta(k,m)-\gamma_b(m)\gamma_b(k)^*$$

This is in agreement with the CAR

$$[b(m),b(k)^*]_+ = \delta(m,k)$$

provided that we assume the CAR

$$[\gamma_b(m),\gamma_b(k)^*]_+ = \delta(m,k)$$

and likewise,

$$[\gamma_c(m),\gamma_c(k)^*]_+ = \delta(m,k)$$

By imposing such restrictions, we can calculate easily the moments of the current density field and hence of the radiated field is a state that is jointly coherent for the Bosons (ie photons) and for the Fermions. We observe that the perturbation to the current density of the Dirac field caused by the interactions between the electron-positron field and the photon field is given upto first order in the photon field and second order in the Fermion field by an expression of the form

$$\delta J^{\mu} = \psi^{(0)*}\alpha^{\mu}\psi^{(1)} + \psi^{(1)*}\alpha^{\mu}\psi^{(0)}$$

$$= -e\psi^{(0)*}(x)\alpha^{\mu}\int S_e(x-y)\gamma^{\mu}A_{\mu}(y)\psi^{(0)}(y)d^4y$$

$$+h.c$$

This expression is manifestly trilinear in the operators. Specifically, it is quadratic in the electron-positron field and linear in the photon field, totally yielding a trilinear term. It can be expressed as

$$\delta J^{\mu}(x) =$$

$$\sum_{k,m,q}(F_1^{\mu}(x|k,m,q)b(k)^*b(m) + F_2^{\mu}(x|k,m,q)b(k)^*c(m)^* +$$

$$F_3^{\mu}(x|k,m,q)c(k)b(m) + F_4(x|k,m,q))c(k)c(m)^*)a(q) + h.c)$$

It should be noted that the photon operatrors $a(q), a(q)^*$ commute with all the electron-positron operators $b(k), b(k)^*, c(k), c(k)^*$. From this expression, it is clear that the photon operators tend to couple the other modes of the electron-positron field and hence produce additional terms in the far field radiation pattern. If we have a state $|\psi>$ of the electron-positron-photon field in which there are $n_e(k) = 0,1$ electrons with momentum-spin index k, $n_p(k) = 0,1$ positrons with momentum-spin index k and $n_{ph}(k) = 0,1,2,...$ photons with momentum-helicity index k for $k = 1,2,...,$ then we can calculate easily the moments of the current fluctuation field $\delta J^{\mu}(x)$ in this state by simply applying the rules

$$b(k)|n_e(k) = 0 >= 0, b(k)|n_e(k) = 1 >= |n_e(k) = 0 >,$$

$$c(k)|n_p(k) = 0 >= 0, c(k)|n_p(k) = 1 >= |n_p(k) = 0 >,$$

$$b(k)^*|n_e(k) = 0 >= |n_e(k) = 1 >, b(k)^*|n_e(k) = 1 >= 0,$$

$$c(k)^*|n_p(k) = 0 >= |n_p(k) = 1 >, c(k)^*|n_p(k) = 1 >= 0,$$

in view of the Pauli-exclusion principle. and likewise for the photon number states

$$a(k)|n_{ph}(k) >= \sqrt{n_{ph}(k)}|n_{ph}(k) - 1 >,$$

$$a(k)^*|n_{ph}(k) >= \sqrt{n_{ph}(k) + 1}|n_{ph}(k) + 1 >$$

More precisely, we can evaluate the moments

$$< \psi|\delta J^{\mu_1}(x_1) \otimes ... \otimes \delta J^{\mu_m}(x_m)|\psi >$$

by noting that the quantity

$$\delta J^{\mu_1}(x_1) \otimes ... \otimes \delta J^{\mu_m}(x_m)$$

is a homogeneous polynomial of degree $3m$ in the electron-positron-photon operators with the photon operators appearing with a total degree of m and the electron-positron operators appearing with a total degree of $2m$. We can also evaluate the above moment in a joint coherent state $|\phi_{ep}(\gamma) \otimes \phi_{ph}(u) >$.

The reference for the material on Fermionic coherent state has been taken from the Master's thesis of Greplova on "Fermionic Gaussian states".

Chapter 4

Supersymmetric Yang-Mills Theory for non-Abelian Gauge Fields

[1] Definition of the universal enveloping algebra of a Lie algebra: Let $\mathfrak{g}$ be a Lie algebra and let $(\mathcal{C}, \pi)$ be a pair such that (a) $\mathcal{C}$ is an associative algebra, (b) $\pi : \mathfrak{g} \to \mathcal{C}$ is a linear mapping satisfying $\pi([X,Y]) = \pi(X)\pi(Y) - \pi(Y)\pi(X) \forall X, Y \in \mathfrak{g}$, (c) $\pi(\mathfrak{g})$ generates $\mathcal{C}$ And (d) if $\mathfrak{U}$ is any associative algebra and $\xi : \mathfrak{g} \to \mathfrak{U}$ is a linear map satisfying $\xi([X,Y]) = \xi(X)\xi(Y) - \xi(Y)\xi(X) \forall X, Y \in \mathfrak{g}$, then there exists an algebra homomorphism $\xi' : \mathcal{C} \to \mathfrak{U}$ such that $\xi'(\pi(X)) = \xi(X) \forall X \in \mathfrak{g}$. Then, $(\mathcal{C}, \pi)$ is called a universal enveloping algebra of $\mathfrak{g}$. Theorem: If $(\pi_k, \mathcal{C}_k), k = 1, 2$ are two universal enveloping algebras of a Lie algebra $\mathfrak{g}$, then they are isomorphic in the sense that there exists an algebra isomorphism $\xi : \mathcal{C}_1 \to \mathcal{C}_2$ such that $\xi(\pi_1(X)) = \pi_2(X) \forall X \in \mathfrak{g}$.

[2] Definition of a super Lie group and a super Lie algebra.

Example: Consider a super Lie group parameterized by just one Bosonic coordinate x and one Fermionic coordinate θ. Thus, (x, θ) is a point in this super Lie group. The composition law for this super Lie group is given by

$$(x_1, \theta_1).(x_2, \theta_2) = (x_1 + x_2 + \gamma\theta_1\theta_2, \theta_1 + \theta_2)$$

where γ is a real scalar. Note that $\theta_1\theta_2$ is Bosonic since it commutes with both Bosonic and Fermionic variables. Let $f(x, \theta)$ be a function on this super Lie group. Then,

$$\frac{\partial}{\partial x_2} f((x_1, \theta_1).(x_2, \theta_2))|_{x_2=0, \theta_2=0} =$$

$$\frac{\partial}{\partial x_1} f(x_1, \theta_1)$$

and

$$\frac{\partial}{\partial\theta_2}f((x_1,\theta_1).(x_2,\theta_2))|_{x_2=0,\theta_2=0}$$

$$(\gamma\theta_1\frac{\partial}{\partial x_1}+\frac{\partial}{\partial\theta_1})f(x_1,\theta_1)$$

This prompts to define the basic Bosonic and Fermionic left invariant vector fields on the super Lie group as

$$\frac{\partial}{\partial x},\gamma\theta\frac{\partial}{\partial x}+\frac{\partial}{\partial\theta}$$

More generally, we can consider a super Lie group with n Bosonic and n Fermionic coordinates. Then, a point in this super Lie group is specified as (x,θ) where $x=(x_1,...,x_n)$ and $\theta=(\theta_1,...,\theta_n)$. The composition law is specified as

$$(x^\mu,\theta^a).(x^{'\mu},\theta^{'a})=(x^\mu+x^{'\mu}+\theta^{'T}\Gamma^\mu\theta,\theta^a+\theta^{'a})$$

where Γ^μ is a skew-symmetric matrix. In shorthand notation, we may write this as

$$(x,\theta).(x',\theta')=(x+x'+\theta^T\Gamma\theta,\theta+\theta')$$

Then the basic Bosonic and Fermionic left invariant vector fields on this super Lie group are defined as

$$\frac{\partial}{\partial x'^\mu}f((x,\theta).(x',\theta'))|_{x'=0,\theta'=0}=$$

$$=\frac{\partial f(x,\theta)}{\partial x^\mu}$$

and

$$\frac{\partial}{\partial\theta'^a}f((x,\theta).(x',\theta'))|_{x'=0,\theta'=0}$$

$$=(\Gamma^\mu\theta)_a\frac{\partial}{\partial x^\mu}+\frac{\partial}{\partial\theta^a}$$

or equivalently in vector notation, the basic Bosonic and Fermionic left invariant vector fields on the super Lie group are given by

$$\frac{\partial}{\partial x^\mu},\Gamma^\mu\theta.\frac{\partial}{\partial x^\mu}+\frac{\partial}{\partial\theta}$$

This formalism was in the specialized context of four Bosonic space-time and four Fermionic space-time dimensions first proposed by Salam and Stratdhee as representations of the supersymmetric Lie algebra with the aim to perform computations on super fields.

References:

[1] Steven Weinberg, "The quantum theory of fields, vol.III, Supersymmetry", Cambridge University Press.

[2] V.S.Varadarajan, "Supersymmetry for Mathematicians, An Introduction", Courant Institute Lecture Notes.

[3] Supersymmetric Yang-Mills theory in the absence of scalar,Dirac, gravitational and auxiliary fields. The Lagrangian for this theory is

$$L = F^A_{\mu\nu}F^{\mu\nu A} + c_1\bar{\lambda}^A\gamma^\mu D_\mu\lambda^A$$

where

$$F^A_{\mu\nu} = V^A_{\nu,\mu} - V^A_{\mu,\nu} + C(ABC)V^\mu_B V^\nu_C$$

with V^μ_A the Bosonic non-Abelian gauge fields and λ^A the Majorana non-Abelian gaugino fields. D_μ is the gauge covariant derivative which acts on the gaugino fields in the adjoint representation:

$$D_\mu\lambda^A = \partial_\mu\lambda^A + C(ABC)V^B_\mu\lambda^C$$

The supersymmetry transformation is given by

$$\delta\lambda^A = \gamma^{\mu\nu}\alpha.F^A_{\mu\nu} = \gamma^{\mu\nu}\alpha.F^A_{\mu\nu}$$

$$\delta V^A_\mu = \bar{\alpha}\gamma_\mu\chi^A$$

so that

$$\delta F^A_{\mu\nu} = \bar{\alpha}(\gamma_\nu\chi^A_{,\mu} - \gamma_\mu\chi^A_{,\nu})$$

Note that

$$\delta F^A_{\mu\nu}$$

will contain another "nonlinear term"

$$C(ABC)\delta(V^B_\mu V^C_\nu) =$$

$$C(ABC)(\delta V^B_\mu.V^C_\nu + V^B_\mu\delta V^C_\nu) =$$

$$\bar{\alpha}.C(ABC)(V^C_\nu\gamma_\mu\chi^B + V^B_\mu\gamma_\nu\chi^C)$$

This quantity is antisymmetric in (μ,ν) since $C(ABC)$ is antisymmetric w.r.t interchange of B,C while the term within the brackets (.) is symmetric w.r.t interchange of B,C. In order to check supersymmetry invariance, we can adopt any gauge since the Lagrangian is gauge invariant. Specifically, we can choose a point X, as Weinberg says and a gauge such that $V^A_\mu(X) = 0$ and then verify supersymmetry invariance at X. It is for this reason, that the nonlinear terms have been ignored in the supersymmetry transformation of $F^A_{\mu\nu}$. It should be noted that a gauge transformation does not depend upon the partial derivatives of the fields, ie, under an infinitesimal gauge transformation defined by Λ^A, we have

$$\delta V^A_\mu = \Lambda^A_{,\mu} + C(ABC)\Lambda^B V^C_\mu$$

$$\delta\lambda^A = C(ABC)\Lambda^B\lambda^C$$

which shows that if $V^A_\mu(X) \neq 0$, then we can apply a sequence of such infinitesimal gauge transformations so that $V^A_\mu(X) = 0$. Another way to check this is as follows: Let $g(x)$ denote an element of the gauge group. Then under a finite gauge transfromation defined by this element, $V_\mu = V_\mu^A t_A$ changes at X to

$$g(X)V_\mu(X)g(X)^{-1} + (\partial_\mu g(X))g(X)^{-1}$$

which can be made zero by choosing $g(X)$ so that

$$V_\mu(X) = -g(X)^{-1}(\partial_\mu g(X))$$

Now once we have made $V_\mu(X)$ equal to zero via this gauge transformation, we can check supersymmetry at X and then apply the inverse gauge transformation to deduce supersymmetry at all the gauges. Another way to see this is to first apply a supersymmetry transformation to the Lagrangian, then apply the above gauge transformation to this resultant Lagrangian so that all terms involving $V_\mu^A(X)$ become zero with the supersymmetrically transformed Lagrangian remaining invariant. This resulting Lagrangian could equivalently be derived from the original Lagrangian by applying the supersymmetry transformation with $V_\mu^A(X) = 0$ and hence if we are able to verify that this Lagrangian coincides with the one with which we started except for a pure derivative term, we would then be done.

Let us look at some consequences of this super-Yang-Mills Lagrangian density. Specifically, setting the variational derivative w.r.t V_μ^A to zero gives

$$D_\nu F^{\mu\nu A} + c_1 C(BAC)\bar{\lambda}^B \gamma^\mu \lambda^C = 0,$$

and setting the variational derivative of the same Lagrangian density w.r.t $\bar{\lambda}^A$ to zero gives us

$$\gamma^\mu D_\mu \lambda^A = 0$$

or equivalently,

$$\gamma^\mu(\partial_\mu \lambda^A + C(ABC)V_\mu^B \lambda^C) = 0$$

which is just like the massless Dirac equation but with the covariant derivative acting in the adjoint representation. Recall that the massless Dirac equation is given by

$$\gamma^\mu(\partial_\mu \psi - iV_\mu^B t_B \psi) = 0$$

where $\{t_B\}$ are the Hermitian generators of the gauge group. We can always choose a basis so that

$$-i(t_B)_{AC} = C(ABC)$$

and then our Dirac equation in this basis for ψ coincides with the differential equation satisfied by the gaugino field λ.

Remark: In case gravity is also taken into account, we would require local supersymmetry invariance which is a more difficult matter. In that case, all the component fields have to be brought in in order to obtain invariance under diffeomorphisms, local supersymmetry, gauge and local Lorentz transformations.

[4] **Construction of the supersymmetric Lagrangian for non-Abelian gauge fields.**

Any superfield can be brought to the form where the only non-vanishing components are V_μ^A, λ^A, D^A via an extended gauge transformation. Such an extended gauge transformation is given by

$$\Gamma(x,\theta) \to exp(it_A\Omega^A).\Gamma(x,\theta).exp(-it^A\Omega_A^*)$$

where t^A are the Hermitian generators of the gauge group and Ω^A are arbitrary left Chiral superfields. Here,

$$\Gamma(x,\theta) = exp(t^A V_A(x,\theta))$$

and in the Wess-Zumino gauge, the gauge superfield is given by

$$V = V^A t_A = t.V$$

where

$$\begin{aligned} V^A &= V_\mu^A \theta^T \epsilon \gamma^\mu \theta + \theta^T \epsilon \theta . \theta^T \gamma^5 \epsilon \lambda^A + (\theta^T \epsilon \theta)^2 D^A \\ &= 2V_\mu^A \theta_R^T \epsilon \gamma^\mu \theta_L + (\theta_R^T \epsilon \theta_R \theta_L^T \\ &\quad + \theta_L^T \epsilon \theta_L . \theta_R^T) \gamma^5 \epsilon \lambda^A \\ &\quad + (\theta_R^T \epsilon \theta_R)(\theta_L^T \epsilon \theta_L) D^A \end{aligned}$$

Once V has been brought into the Wess-Zumino gauge, it is does not remain this gauge under an arbitrary supersymmetric transformation but after applying an arbitrary supersymmetric transformation, it can be once again be brought to the Wess-Zumino gauge by an extended gauge transformation. Moreover, the action constructed from the gauge field V^A in the Wess-Zumino gauge is neither invariant under an arbitrary supersymmetry transformation nor under an extended gauge transformation but it is invariant under ordinary gauge transformations and also under an application of a general supersymmetry transformation followed by an extended gauge transformation that brings it back to the Wess-Zumino gauge. This gauge action is also invariant under ordinary gauge transformations. In order to deal with this problem, we first construct out of the gauge field V^A in the Wess-Zumino gauge spinor superfield W_L that is left Chiral and then observe that $Tr(W_L^T \epsilon W_L)$ is again left Chiral and invariant under extended gauge transformations since under extended gauge transformations, W_L transforms to $exp(it_A\Omega^A).W_L.exp(-it_A\Omega^A)$. Since then W_L is left Chiral, its "F-part", ie, the coefficient of $\theta_L^T \epsilon \theta_L$ is a supersymmetric and gauge invariant Lagrangian and hence so are its real and imaginary parts. Thus, using these real and imaginary parts, we are able to get a supersymmetric gauge action that generalizes the Lagrangian of the Yang-Mills gauge field in conventional non-supersymmetric quantum field theory. The real part of $[Tr(W_L^T \epsilon W_L)]_F$ has a term of the form $F_{\mu\nu}^A F^{\mu\nu A}$ that is the Yang-Mills Lagrangian ($F_{\mu\nu}^A = [D_\mu, D_\nu]$ with D_μ the covariant derivative w.r.t the connection V_μ^A) while its imaginary part has a term $\epsilon^{\mu\nu\rho\sigma} F_{\mu\nu}^A F_{\rho\sigma}^A$ that is responsible for anomalies in conventional quantum field theory and its integral over space-time is simply the index of the Dirac operator. Apart from these two terms, there is a gaugino term involving λ^A in the Lagrangian which is just like the massless Dirac Lagrangian with the exception that we use the covariant derivative of λ^A w.r.t. the connection V_μ^A, ie, the covariant derivative on λ^A acts in the adjoint representation unlike the Dirac Lagrangian in which the covariant derivative of the Dirac wave function acts in the usual gauge group vector representation. The reason for this difference is that the Dirac wave function represents a matter field while the gaugino λ^A-field represents a part of the gauge superfield. There is also an auxiliary D^A-term in the gauge Lagrangian. It should be noted that apart from the gauge Lagrangian, the matter Lagrangian of the Chiral field is modified from $[\Phi^*\Phi]_D$ to $[\Phi^*\Gamma\Phi]_D$ which is again supersymmetric and gauge invariant and is the supersymmetric analogue of the Yang-Mills matter field Lagrangian interacting with the non-Abelian gauge field or in the Abelian case, the Dirac matter field Lagrangian interacting with the electromagnetic field.

The spinor left Chiral gauge superfield is given by

$$W_L = D_R^T \epsilon D_R exp(-t.V) D_L exp(t.V)$$

If Ω is a left Chiral superfield, then the transformation law of the gauge superifeld V is given by

$$exp(t.V) \to exp(i\Omega^*) exp(t.V).exp(-i\Omega)$$

Taking inverse gives us

$$exp(-t.V) \to exp(i\Omega) exp(-t.V).exp(-i\Omega^*)$$

Then, it is easily seen that under this generalized gauge transformation, W_L transforms as

$$W_L \to exp(i\Omega) W_L.exp(-i\Omega)$$

which is indeed left Chiral as it should be. Here

$$\Omega = t_A \Omega^A(x_+, \theta_L)$$

is a matrix left Chiral superfield. To prove this result, we must use the fact that D_L annihilates $exp(i\Omega^*)$ since Ω^* is right Chiral and that $D_R^T \epsilon D_R D_L exp(-i\Omega) = 0$ since $D_R.exp(-i\Omega) = 0$ and $[D_R^T \epsilon D_R, D_L]$ equals a Bosonic differential operator times D_R since $[D_R, D_L]$ is a Bosonic differential operator with constant coefficients.

Thus,

$$Tr[W_L^T \epsilon W_L]$$

is invariant under extended gauge transformations and the real and imaginary parts of its F-components are thus both invariant under extended gauge transformations and under supersymmetry as also under Lorentz transformations.

Under a restricted gauge transformation defined by

$$\Omega = \Lambda(x_+) = t_A \Lambda^A(x_+)$$

we see easily that these facts specialize to

$$W_L^T \epsilon W_L \to exp(i\Lambda(x_+)).W_L^T \epsilon W_L.exp(-i\Lambda(x_+))$$

and hence

$$[W_L^T \epsilon.W_L]_F \to exp(i\Lambda(x_+))[W_L^T \epsilon W_L]_F.exp(-i\Lambda(x_+))$$

where

$$\Lambda(x) = t_A \Lambda^A(x)$$

is a Bosonic matrix field. Then

$$Tr([W_L^T \epsilon._L]_F) \to Tr([W_L^T \epsilon W_L]_F)$$

which proves the restricted gauge invariance of the gauge Lagrangian

$$L_G = Tr[W_L^T \epsilon W_L]_F$$

Now, we propose to evaluate W_L and then L_G in a restricted gauge for which at a given point X, we have $V_\mu^A(X) = 0$.

$$W_L = D_R^T \epsilon D_R exp(-t.V) D_L exp(t.V) =$$

$$D_R^T \epsilon.D_R(1 - t.V + (t.V)^2/2) D_L(t.V + (t.V)^2/2)$$

$$= D_R^T \epsilon.D_R(D_L(t.V) + D_L((t.V)^2)/2 - (t.V).D_L(t.V))$$

$$= D_R^T \epsilon D_R D_L(t.V) + D_R^T \epsilon.D_R((1/2)(D_L(t.V))(t.V) + ((-1)^{p(V)}/2)t.V D_L(t.V)$$

$$-(t.V) D_L(t.V))$$

Now,

$$D_L(t.V) = t^A D_L(V^A)$$

$$D_L(V^A) = (\gamma^5 \epsilon \partial_{\theta_L} - \gamma^\mu \theta_R \partial_\mu)[2V_\mu^A \theta_R^T \epsilon \gamma^\mu \theta_L + (\theta_R^T \epsilon \theta_R \theta_L^T$$

$$+\theta_L^T \epsilon \theta_L.\theta_R^T)\gamma^5 \epsilon \lambda^A$$

$$+(\theta_R^T \epsilon \theta_R)(\theta_L^T \epsilon \theta_L) D^A]$$

$$= [-2\gamma^5 \gamma^\mu \theta_R.V_\mu^A - (\theta_R^T \epsilon \theta_R).\gamma^5 \epsilon.\lambda^A$$

$$-2\gamma^5 \theta_L \theta_R^T \gamma^5 \epsilon \lambda^A - 2\theta_R^T \epsilon \theta_R \gamma^5 \theta_L.D^A]$$

$$-2V_{\mu,\nu}^A \gamma^\nu \theta_R.\theta_R^T \epsilon \gamma^\mu \theta_L$$

$$-\theta_L^T \epsilon \theta_L.\gamma^\mu \theta_R \theta_R^T \gamma^5 \epsilon \lambda_{,\mu}^A$$

Exercise: [1] Evaluate the other terms to get explicit formulas for the real and imaginary parts of $Tr[W_L^T \epsilon W_L]_F$ and explicitly show that a linear combination of its real and imaginary parts consists of the terms $F^A_{\mu\nu}F^{\mu\nu A}$, $\epsilon^{\mu\nu\rho\sigma}F^A_{\mu\nu}F^A_{\rho\sigma}$, $(\lambda^A)^T\gamma^5\epsilon\gamma^\mu D_\mu\lambda^A$ and $D^A D^A$ where

$$D_\mu\lambda^A = \partial_\mu\lambda^A + C(ABC)V^B_\mu\lambda^C$$

is the covariant derivative of λ^A in the adjoint representation of the gauge group. Explain this result in the context of a supersymmetric generalization of the electromagnetic field Lagrangian or more generally in the context of the non-supersymetric Yang-Mills gauge field Lagrangian with anomalous terms.

[2] Evaluate the path integral of this action using perturbation theory and using this, compute the propagators of the gauge field V^A_μ, the gaugino field λ^A and the auxiliary field D^A.

[3] Write down the field equations for V^A_μ, the gaugino field λ^A noting that the field equations for the auxiliary field D^A which has the field equation $D^A = 0$. Obtain approximate solutions to these field equations using perturbation theory upto the second order in the perturbation parameter δ attached to the nonlinear term in $F^A_{\mu\nu}$:

$$F^A_{\mu\nu} = V^A_{\nu,\mu} - V^A_{\mu,\nu} + \delta.C(ABC)V^B_\mu V^C_\nu$$

Chapter 5

Supersymmetric Quantum Stochastic Filtering Theory

[0].Introduction: It is now a well known fact that corresponding to the Hudson-Parthasarathy-Schrodinger quantum stochastic differential equation that describes the unitary evolution of a quantum system coupled to a noisy bath admits a filtering theory as first proposed by V.P.Belavkin and perfected by John Gough, Kostler and Luc Bouten. The idea here is that the noise in the HPS-qsde is Bosonic comprising differentials of the creation, annihilation and conservation operator valued processes that satisfy the quantum Ito formula or equivalently satisfy the canonical commutation relations for Bosonic fields. Belavkin first constructed non-demolition measurements for such systems using which one could construct the conditional expectation of a Heisenberg observable evolving in the HPS-qsde framework and that this conditional expectation could be computed on a real time basis leading to the Belavkin filter that generalizes the classical Kushner-Kallianpur filter in stochastic filtering theory. The setup of the HPS-qsde presumes that the noisy bath is Bosonic, ie, consists of photons and that the system can be anything. However, if the bath is Fermionic, for example, if it consists of electrons and positrons, then the noise differentials appearing in the qsde should be Fermionic noise differentials that satisfy the canonical anticommutation relations. In fact, Hudson and Parthasarathy have developed the calculus of Fermionic noise using Bosonic noise differentials modulated by -1 raised to the power of a conservation process. To construct a non-demolition measurements for qsde's driven by Fermionic noise, we require that the Fermionic noise differentials commute at two different time instants which is not true. In fact they anticommute. To rectify this problem, we multiply the Fermionic noise processes appearing in the qsde as well as in the non-demolition measurement by Grassmannian/Fermionic parameters that mutually anticommute and hence restore commutativity of the Fermionic noise differentials. The filtering has then to be carried out in Fermionic coherent states. These are states parametrized by Grassmannian parameters which mutually anticommute

and also anticommute with the Fermionic operators. Physical meaning to the filtering results in such coherent states can be obtained only after averaging the results by means of a weight function dependent upon the Grassmannian parameters using the Berezin integral. This amounts to choosing our state as a weighted linear combination of Fermionic coherent states. This idea also plays a fundamental role in the theory of quantum antennas wherein the electromagnetic field produced by the current density generated by electrons and positrons in the Dirac picture is a quadratic combination of the Fermionic creation and annihilation operators and hence by applying the retarded potential formula, the electromagnetic field radiated out by such a quantum antenna is also a quadratic functional of the Fermionic creation and annihilation operators. Computing the quantum statistical moments of the radiation field in a Fermionic coherent state results in a function of the Grassmannian parameters which after a weighted Berezin integral approximation gives real and complex numbers which can be interpreted physically in terms of correlations, energy, power and phase.

[1] The Fermionic noise here is of the form

$$dJ(t) = (-1)^{\Lambda(t)}dA(t), dJ^*(t) = (-1)^{\Lambda(t)}dA^*(t)$$

These satisfy the CAR

$$\{J(t), J(s)^*\} = min(t,s), \{J(t), J(s)\} = 0$$

just as the Bosonic noise here satisfies the CCR

$$[A(t), A(s)^*] = min(t,s), [A(t), A(s)] = 0$$

In a quantum stochastic dynamical system with Fermionic noise, the qsde for the unitary evolution operator will be given by

$$dU(t) = (-(iH + P)dt + LdJ(t) - L^*dJ(t)^*)U(t), P = LL^*/2$$

More generally, let X be any finite Hermitian matrix. We define

$$d\xi_k(t) = (-1)^{\lambda(X_t)}dA_k(t)$$

where

$$X_t = X\chi_{[0,t]}$$

Then,

$$d\xi_k^*(t) = (-1)^{\lambda(X_t)}dA_k(t)^*$$

We have

$$(-1)^{\lambda(X_t)}|e(u)>= exp(i\pi\lambda(X_t))|e(u)>=$$

$$\Gamma(exp(i\pi X_t))|e(u)>= |e(exp(i\pi X_t)u)>=$$

$$|e(exp(i\pi X)u\chi_{[0,t]} + u\chi_{(t,\infty)})>$$

$$d\xi_k(t)|e(u)>= exp(i\lambda(\pi X\chi_{[0,t]}))dA_k(t)|e(u)>=$$

$$u_k(t)dt.|e(exp(i\pi X\chi_{[0,t]})u)>=u_k(t)dt|exp(i\pi X)u\chi_{[0,t]}+u\chi_{(t,\infty)}>$$
$$=u_k(t)dt|(-1)^X u\chi_{[0,t]}+u\chi_{(t,\infty)}>$$

We now compute for $s<t$ the quantities

$$dA_k(s).(-1)^{\lambda(X_t)}$$

and

$$(-1)^{\lambda(X_t)}dA_k(s)$$

by looking at their action on $|e(u)>$. The first gives

$$dA_k(s)(-1)^{\lambda(X_t)}|e(u)>=$$

$$dA_k(s)|e(Hu\chi_{(0,t]}+u\chi_{(t,\infty)})>$$
$$=(Hu)_k(s)ds|e(Hu\chi_{[0,t]}+\chi_{(t,\infty)})>$$

while the second gives

$$(-1)^{\lambda(X_t)}dA_k(s)|e(u)>=$$

$$u_k(s)ds|e(Hu\chi_{[0,t]}+\chi_{(t,\infty)})>$$

where

$$H=(-1)^X=exp(i\pi.X)$$

In particular, assuming that $u\in\mathbb{C}^N$ so that the Boson Fock space under consideration is $\Gamma(L^2(\mathbb{R}_+)\otimes\mathbb{C}^N)$ and then taking

$$X=diag[0_r,I_{n-r}]$$

gives us

$$H=diag[I_r,-I_{n-r}]$$

and then if $1\leq k\leq r$, we get that

$$(Hu)_k=u_k,$$

while if $r+1\leq k\leq N$, we get that

$$(Hu)_k=-u_k$$

so that we get the following super-commutation relations:

$$(-1)^{\lambda(X_t)}dA_k(s)=(-1)^{\sigma(k)}dA_k(s)(-1)^{\lambda(X_t)},s\leq t$$

where $\sigma(k)=1$ if $1\leq k\leq r$ and $\sigma(k)=-1$ if $r+1\leq k\leq N$. Taking the adjoint of this equation gives us

$$dA_k(s)^*.(-1)^{\lambda(X_t)}=(-1)^{\sigma(k)}dA_k(s)^*(-1)^{\lambda(X_t)},s\leq t$$

Now define

$$J_k(t)=\int_0^t(-1)^{\lambda(X_s)}dA_k(s),$$

so that

$$J_k(t)^* = \int_0^t (-1)^{\lambda(X_s)} dA_k(s)^*$$

Then, have for $s < t$,

$$dJ_k(s)dJ_m(t) = (-1)^{\lambda(X_s)} dA_k(s)(-1)^{\lambda(X_t)} dA_m(t)$$

$$= (-1)^{\sigma(k)}(-1)^{\lambda(X_s)+\lambda(X_t)} dA_k(s)dA_m(t)$$

from which we easily deduce that

$$dJ_k(s).dJ_m(t) = (-1)^{\sigma(k)} dJ_m(t).dJ_k(s), s < t$$

and hence

$$J_k(t)J_m(s) - (-1)^{\sigma(k)} J_m(t)J_k(s) = 0, \forall s \le t \in \mathbb{R}_+, k, m = 1, 2, ..., N$$

Again, for $s < t$,

$$dJ_k(t).dJ_m(s)^* = (-1)^{\lambda(X_t)} dA_k(t)(-1)^{\lambda(X_s)} dA_m(s)^*$$

$$= (-1)^{\sigma(m)}(-1)^{\lambda(X_s)} dA_m(s)^*(-1)^{\lambda(X_t)} dA_k(t)$$

$$= (-1)^{\sigma(m)} dJ_m(s)^* dJ_k(t)$$

where we have used the fact that $dA_k(t)$ and $\lambda(X_t)$ commute with $\lambda(X_s)$ and $dA_m(s)^*$ commutes with $dA_k(t)$ for $s < t$. Further,

$$dJ_k(t).dJ_m(t)^* = dt.\delta_{km}$$

since $dA_k(t)dA_m(t)^* = dt.\delta_{km}$ by the Hudson-Parthasarathy quantum Ito formula and $dA_k(t), dA_k(t)^*$ both commute with $\lambda(X_t)$ and we have also used the fact that $\lambda(X_t)$ has integer eigenvalues since by quantum Ito's formula, $(d\lambda(X_t))^2 = d\lambda(X_t)$ and hence $d\lambda(X_t)$ has only zero and one as its eigenvalues. Further, we also have by quantum Ito's formula,

$$dJ_m(t)^* dJ_k(t) = 0$$

Thus, we obtain on integration,

$$dJ_k(t)J_m(s)^* - (-1)^{\sigma(m)} J_m(s)^* dJ_k(t) = 0, s < t,$$

$$dJ_k(t)dJ_m(t)^* - (-1)^{\sigma(m)} dJ_m(t)^* dJ_k(t) = \delta_{km} dt$$

Now, taking the adjoint of the first equation and interchanging k and m gives us

$$J_k(s)dJ_m(t)^* - (-1)^{\sigma(k)} dJ_m(t)^* J_k(s) = 0, s < t$$

from which we infer that

$$J_k(s)J_m(t)^* - (-1)^{\sigma(k)} J_m(t)^* J_k(s) = s.\delta_{km}, s \le t$$

[2] Supersymmetric filtering theory in quantum stochastic calculus:

$$dJ(t) = (-1)^{\Lambda(t)}dA(t), dJ(t)^* = (-1)^{\Lambda(t)}dA(t)^*$$

Then,

$$[A(t), A(s)^*] = min(t,s), \{J(t), J(s)^*\} = min(t,s)$$

Consider a Fermionic qsde

$$dU(t) = (-(iH + P)dt + LdJ(t) - L^*dJ(t)^*)U(t)$$

Measurement model

$$Y_i(t) = cA(t) + \bar{c}A(t)^*, Y_o(t) = U(t)^*Y_i(t)U(t)$$

Let $t > s$. Then, compute

$$d_t(U(t)^*Y_i(s)U(t)) = dU(t)^*Y_i(s)U(t) + U(t)^*Y_i(s)dU(t) + dU(t)^*Y_i(s)dU(t)$$

This is not zero since $dJ(t), dJ(t)^*$ do not commute with $A(s), A(s)^*$. So $Y_o(t)$ is not a non-demolition measurement. Suppose we take

$$Y_i(t) = cJ(t) + \bar{c}J(t)^*, Y_o(t) = U(t)^*Y_i(t)U(t)$$

Then for $Y_o(t)$ to be non-demolition, we must have that $dJ(t)$ commutes with $J(s), J(s)^*$ for $t > s$. However, it is in fact true that $dJ(t)$ anticommutes with $dJ(s), dJ(s)^*$ for $t > s$ and hence also with $J(s), J(s)^*$ for $t > s$. So in order to obtain a quantum filtering theory for Fermionic noise, we must introduce anticommuting Fermionic parameters into the dynamics of $U(t)$ as well as into the measurement processes so that the measurement process becomes non-demolition. Specifically, let θ_1, θ_2 be anticommuting Fermionic parameters. Then, replace the Lindblad operator L by $\theta_1 L$ and let $\theta_1^* = \theta_2$. Thus, L^* gets replaced by $\theta_2 L$. Then, introduce a measurement process

$$Y_o(t) = U(t)^*Y_i(t)U(t), Y_i(t) = c\theta_1 J(t) + \bar{\theta}_2 J(t)^*$$

Then, in order that Y_o be a non-demolition measurement process, we require that

$$dU(t)^*Y_i(s)U(t) + U(t)^*Y_i(s)dU(t) + dU(t)^*Y_i(s)dU(t) = 0, t > s$$

This will be the case if $\theta_1 dJ(t)$ and $\theta_2 dJ(t)^*$ commute with $\theta_1 J(s)$ and $\theta_2 J(s)^*$ for $t > s$ which is true since $dJ(t)$ and $dJ(t)^*$ anticommute with $J(s), J(s)^*$ for $t > s$.

Reference: Timothy Eyre, "Quantum Stochastic Calculus and Representations of Lie Superalgebras", Springer Lecture notes in Mathematics.

[3] More generally, we can consider the qsde

$$dU(t) = L_b^a d\xi_a^b(t)U(t)$$

where L_b^a are system space Lindblad operators and

$$d\xi_a^b(t) = G(t)^{\sigma_b^a} d\Lambda_b^a(t)$$

where $\Lambda_b^a(t)$ satisfy the quantum Ito formula

$$d\Lambda_b^a.d\Lambda_d^c = \epsilon_d^a d\Lambda_b^c(t)$$

where ϵ_b^a is one if $a = b > 0$ and zero otherwise. Here,

$$G(t) = (-1)^{\lambda(H\chi_{[0,t]})}$$

where

$$H = diag[0_r, I_{N-r}]$$

or equivalently,

$$G(t) = \Gamma(exp(i\pi.H\chi_{[0,t]}))$$

and $((\sigma_b^a))$ is an $N \times N$ matrix such that σ_b^a is unity if $1 \leq a, b \leq r$ or $r \leq a, b \leq N$ and -1 otherwise, ie σ is the grading matrix. Then, it is well known (see Timothy Eyre, "Quantum Stochastic Calculus and Representations of Lie Superalgebras", Springer Lecture notes in Mathematics) that

$$\xi_b^a(t).\xi_d^c(s) - (-1)^{\sigma_b^a \sigma_d^c} \xi_d^c(s)\xi_b^a(t) = (\epsilon_d^a \xi_b^c(u) - (-1)^{\sigma_b^a \sigma_d^c} \epsilon_b^c \xi_d^a(u))$$

where

$$u = min(t, s)$$

Note that this super-commutation relation can be expressed in an alternate way as follows:

$$X_a^b Y_c^d \xi_b^a(t)\xi_d^c(s) - (-1)^{\sigma_b^a \sigma_d^c} Y_c^d X_a^b \xi_d^c(s)\xi_b^a(t)$$

$$= \xi_X(t)\xi_Y(s) - (-1)^{p(\xi_X(t))p(\xi_Y(s))} \xi_Y(s)\xi_X(t) =$$

$$\{\xi_X(t), \xi_Y(s)\}_S =$$

$$X_a^b Y_c^d (\epsilon_d^a \xi_b^c(u) - (-1)^{\sigma_b^a \sigma_d^c} \epsilon_b^c \xi_d^a(u))$$

$$= (X_a^b \epsilon_d^a Y_c^d)\xi_b^c(u) - ((-1)^{\sigma_b^a \sigma_d^c} Y_c^d \epsilon_b^c X_a^b)\xi_d^a(u)$$

$$= (X\epsilon.Y)_c^b \xi_b^c(u) - [(-1)^{\sigma(X)\sigma(Y)}(Y\epsilon.X)]_a^d \xi_d^a(u)$$

$$= \xi_{\{X,Y\}_S}$$

where

$$\{X, Y\}_S = X.\epsilon.Y - (-1)^{\sigma(X)\sigma(Y)} Y\epsilon.X$$

is the super Lie bracket between the matrices X and Y with the weight factor ϵ.

We can choose the Lindblad operators L_b^a such that $U(t)$ is unitary for every t. Now suppose we address the filtering theory problem in this generalized quantum noise context. Our measurement model is

$$Y_o(t) = U(t)^* Y_i(t) U(t), Y_i(t) = c_\alpha^\beta \xi_\beta^\alpha(t)$$

where c_α^β are complex numbers with

$$\bar{c}_\alpha^\beta = c_\beta^\alpha$$

in order to guarantee that Y_i, Y_o are Hermitian. Then, for Y_o to be non-demolition we require that $d\xi_\beta^\alpha(s)$ commute with $\xi_\sigma^\rho(t)$ for all $t \geq s$. This is not true in view of the above supercommutation relations. However, it can be made to commmute if we replace the noise process $\xi_\beta^\alpha(t)$ with $\theta_\beta^\alpha.\xi_\beta^\alpha(t)$ where θ_β^α are super-parameters satisfying the supercommutation relations

$$\theta_b^a \theta_d^c - (-1)^{\sigma_b^a \sigma_d^c} \theta_d^c \theta_b^a = 0$$

In this way, one can develop a cogent quantum Belavkin filter. For the final filtering results, one has to average out the filtered output state w.r.t a generalized Berezin integral w.r.t a weight function $\phi(\theta)$ of the super-parameters θ_b^a.

Remark: When instead of Bosonic and Fermionic processes, we consider Bosonic and Fermionic fields, then we get a quantum stochastic superfield theory. For example, in the Boson Fock space $\Gamma_s(\mathcal{H})$, we have the Bosonic creation and annihilation fields $a(u)^*, a(u), u \in \mathcal{H} = L^2(\mathbb{R}_+) \otimes \mathbb{C}^N$ and in Fermionic Fock space, $\Gamma_a(\mathcal{K})$ we have, the Fermionic creation and annihilation fields $J(u)^*, J(u), u \in \mathcal{K}$ where if K is assumed to be finite dimensional, then the action of $J(u), J(u)^*$ in $\Gamma_a(K)$ is defined by

$$J(u)^*(e_1 \wedge ... \wedge e_m) = u \wedge e_1 \wedge ... \wedge e_m,$$

$$J(u)(e_1 \wedge ... \wedge e_m) = \sum_{k=1}^m (-1)^{k-1} < u, e_k > (e_1 \wedge ... \wedge \hat{e}_k \wedge ... \wedge e_m)$$

To get supersymmetry, we must consider the tensor product space

$$\Gamma(H, K) = \Gamma_s(\mathcal{H}) \otimes \Gamma_a(\mathcal{K})$$

[4] Some more remarks on supersymmetry in quantum stochastic calculus. Let

$$d\xi_b^a(t) = G(t)^{\sigma(a,b)} d\Lambda_b^a(t)$$

where

$$G(t) = (-1)^\lambda (X_t)$$

with

$$X_t = X\chi_{[0,t]}, X = diag[0_r, I_{n-r}]$$

Define

$$\sigma(a) = 0, 1 \leq a \leq r, \sigma(a) = 1, r+1 \leq a \leq n, \sigma(a,b) = \sigma(a) + \sigma(b)(mod2)$$

Let

$$H = (-1)^X = diag[I_r, -I_{n-r}]$$

For $s < t$, we have

$$< e(v)|G(t)^{\sigma(a,b)} d\Lambda_d^c(s)|e(u) >=$$

$$< G(t)^{\sigma(a,b)} e(v)|d\Lambda_d^c(s)|e(u) >=$$

$$< dA_d(s).e(H^{\sigma(a,b)} v.\chi_{[0,t]} + v.\chi_{(t,\infty)})|dA_c(s)|e(u) > /ds$$

$$= (H^{\sigma(a,b)} \bar{v}(s))_d u_c(s) ds < e(H^{\sigma(a,b)} v.\chi_{[0,t]} + v.\chi_{(t,\infty)})|e(u) >$$

Now,

$$(H^{\sigma(a,b)} \bar{v}(s))_d = H_{dd}^{\sigma(a,b)} \bar{v}_d(s)$$

$$= (-1)^{\sigma(d)\sigma(a,b)} \bar{v}_d(s)$$

Thus, we get

$$< e(v)|G(t)^{\sigma(a,b)} d\Lambda_d^c(s)|e(u) >$$

$$= (-1)^{\sigma(d)\sigma(a,b)} \bar{v}_d(s) u_c(s) ds. < e(v)|G(t)^{\sigma(a,b)}|e(u) >, s < t$$

On the other hand,

$$< e(v)|d\Lambda_d^c(s).G(t)^{\sigma(a,b)}|e(u) >=$$

$$< d\Lambda_c^d(s) e(v)|G(t)^{\sigma(a,b)}|e(u) >$$

$$=< dA_d(s) e(v)|dA_c(s)|e(H^{\sigma(a,b)} u\chi_{[0,t]} + u\chi_{(t,\infty)}) > /ds$$

$$= \bar{v}_d(s) H_{cc}^{\sigma(a,b)} u_c(s) ds < e(v)|G(t)^{\sigma(a,b)}|e(u) >$$

$$= (-1)^{\sigma(c)\sigma(a,b)} \bar{v}_d(s) u_c(s) ds < e(v)|G(t)^{\sigma(a,b)}|e(u) >$$

Combining these two identities, we deduce that for $s < t$,

$$G(t)^{\sigma(a,b)} d\Lambda_d^c(s) = (-1)^{\sigma(c,d)\sigma(a,b)} d\Lambda_d^c(s).G(t)^{\sigma(a,b)}$$

Thus, we get for $s < t$

$$cG(t)^{\sigma(a,b)} d\Lambda_b^a(t).G(s)^{\sigma(c,d)} d\Lambda_d^c(s)$$

$$-(-1)^{\sigma(c,d).\sigma(a,b)}.G(s)^{\sigma(c,d)} d\Lambda_d^c(s).G(t)^{\sigma(a,b)} d\Lambda_b^a(t) = 0$$

which is the same as

$$d\xi_b^a(t).d\xi_d^c(s) - (-1)^{\sigma(a,b)\sigma(c,d)} d\xi_d^c(s).d\xi_b^a(t) = 0, s < t$$

and moreover,

$$d\xi_b^a(t)d\xi_d^c(t) = G(t)^{\sigma(a,b)+\sigma(c,d)}\epsilon_d^a d\Lambda_b^c(t)$$

and hence

$$d\xi_b^a(t)d\xi_d^c(t) - (-1)^{\sigma(a,b)\sigma(c,d)}d\xi_d^c(t).d\xi_b^a(t)$$

$$= G(t)^{\sigma(a,b)+\sigma(c,d)}\epsilon_d^a d\Lambda_b^c(t)$$

$$-(-1)^{\sigma(a,b)\sigma(c,d)}G(t)^{\sigma(a,b)+\sigma(c,d)}\epsilon_b^c.d\Lambda_d^a(t)$$

$$= G(t)^{\sigma(a,b)+\sigma(c,d)-\sigma(c,b)}\epsilon_d^a d\xi_b^c(t)$$

$$-(-1)^{\sigma(a,b)\sigma(c,d)}G(t)^{\sigma(a,b)+\sigma(c,d)-\sigma(a,d)}\epsilon_b^c.d\xi_d^a(t)$$

Now,

$$\sigma(a,b) + \sigma(c,d) - \sigma(c,b) = \sigma(a) + \sigma(d) = \sigma(a,d)$$

$$\sigma(a,b) + \sigma(c,d) - \sigma(a,d) = \sigma(b) + \sigma(c) = \sigma(b,c)$$

and noting that ϵ_b^a is zero if $a \neq b$, it follows that

$$\epsilon_d^a.G(t)^{\sigma(a,d)} = \epsilon_d^a$$

since $G(t)^m = 1$ if m is even. Thus, we get

$$d\xi_b^a(t)d\xi_d^c(t) - (-1)^{\sigma(a,b)\sigma(c,d)}d\xi_d^c(t).d\xi_b^a(t)$$

$$= \epsilon_d^a d\xi_b^c(t) - (-1)^{\sigma(a,b)\sigma(c,d)}\epsilon_b^c.d\xi_d^a(t)$$

Reference:Timothy Eyre, Quantum stochastic calculus and Lie super-algebras, Springer, Lecture notes in mathematics.

[5] **Large deviation problems in supersymmetry**

Exercises.

[1] A rigid body at time $t = 0$ occupies the spatial volume region B. The body is filled with electrons and positrons and we know from basic quantum electrodynamics, that its Hamiltonian density is given by

$$\mathcal{H} = \psi(x)^*((\alpha, -i\nabla + eA(x)) + \beta m)\psi(x) - e\Phi(x)\psi^*(x)\psi(x)$$

where $\psi(x)$ is the four component Dirac wave function, $A(x)$ is the magnetic vector potential and $\Phi(x)$ is the electric scalar potential. The We can express this Hamiltonian density as

$$\mathcal{H} = \mathcal{H}_0 + eV$$

where

$$\mathcal{H}_0 = \psi^*((\alpha, -i\nabla) + \beta m)$$

is the free Dirac Hamiltonian density and

$$V = \psi^*(x)\alpha_r\psi(x)A_r(x) - e\Phi(x)\psi^*(x)\psi(x)$$

is the interaction Hamiltonian density between the electron-positron field and the photon field. We assume that the electromagnetic field is a zero mean Gaussian field and the exact field equations for the Dirac field are

$$[(\alpha, -i\nabla + eA(x)) + \beta m - e\Phi(x)]\psi(x) = i\partial_t\psi(x)$$

The problem is to determine the LDP rate function for the random wave function $\psi(x)$ in the first quantized scenario as $e \to 0$. Or if X is an observable, ie, in the position representation, $X = ((X(r, r') : r, r' \in \mathbb{R}^3)$ where $X(r, r')$ is 4×4 matrix satisfying the Hermitian property, ie,

$$X(r, r')^* = X(r', r)$$

then to compute the LDP rate function for the family of real random variables

$$< \psi|X|\psi >= \int \psi(t, r)^* X(r, r')\psi(t, r')d^3rd^3r'$$

as $e \to 0$. For this, we must first of all express the solution to the above electromagnetically perturbed Dirac wave function as a Dyson series in the interaction representation, ie, in powers of the electronic charge e.

[2] The simplest super-gravity Lagrangian density is given by

$$L_{SG} = e.[R^{mn}_{\mu\nu} e^{\mu}_m e^{\nu}_n + K.\bar{\chi}_{\mu}\Gamma^{\mu\nu\rho} D_{\nu}\chi_{\rho}]$$

where e^{μ}_m is a usual tetrad field, the spinor connection of the gravitational field is given by

$$a.\omega^{mn}_{\mu}\Gamma_{mn}$$

and the covariant derivative D_{ν} on quantities like the gravitino field χ_{μ} which are spinors having a vector index also is given by (See M.Green, J.Schwarz and E.Witten "Superstring theory") is

$$D_{\nu}\chi_{\rho} = \partial_{\nu}\chi_{\rho} - \Gamma^{\alpha}_{\nu\rho}\chi_{\alpha} + a.\omega^{mn}_{\nu}\Gamma_{mn}\chi_{\rho}$$

We've already seen the supersymmetry transformation under which the action corresponding to this Lagrangian density is invariant. We note that $e = det((e^n_{\mu})) = \sqrt{g}$ where $g = ((g_{\mu\nu}))$. The spinor version of the Riemann curvature tensor $R^{mn}_{\mu\nu}$ is given by

$$R^{mn}_{\mu\nu} = \omega^{mn}_{\nu,\mu} - \omega^{mn}_{\nu,\mu} + [\omega_{\mu}, \omega_{\nu}]^{mn}$$

and the spinor gravitational connection ω^{mn}_{μ} is defined by its field equation, ie, by setting the variational derivative of the supergravity action w.r.t it to zero. We leave it as an exercise to express this spinor connection in terms of the tetrad (graviton) field and the gravitino field χ_{μ}. Note that if the gravitino field were not present, then the spinor connection would be defined by

$$0 = D_{\nu}e^n_{\mu} = e^n_{\mu,\nu} + a.\omega^{nm}_{\nu}e_{m\mu} + \Gamma^{\rho}_{\nu\mu}e^n_{\rho} = 0$$

This equation could also be derived by setting the variation of the above supergravity Lagrangian without the gravitino term w.r.t the spinor connection to zero. We leave this verification as an exercise. Now suppose that we break this supersymmetry by say matter terms interacting with the graviton field and superpartners of the matter field interacting with the gravitino field. Such terms can of course be added with a redefinition of the supersymmetry transformations to ensure that the resulting action after interaction with supermatter still remains supersymmetric. For a nice discussion of this, see Steven Weinberg, "The quantum theory of fields, vol.III, Supersymmetry", Cambridge University Press. However, proceeding with our simplified formalism, we add such random supersymmetry breaking terms and then ask the question, what is the probability that if δS is the change in the action under an infinitesimal supersymmetry transformation, then δS will be greater in magnitude than a given threshold δ_0 ?

Chapter 6

Problems and Study Projects in non-Abelian Gauge and String Theory

[0].Introduction. The notion of a connection on a principal bundle is described in the context of the Yang-mills field equations in curved space-time. A vector bundle on whose fibres a Lie group of transformations act is roughly speaking a fibre bundle. The notion of a connection and a covariant derivative on such a bundle compatible with the Lie group structure is introduced. In general relativity, the base manifold is four dimensional space-time which has a natural connection inherited from the space-time Riemannian metric. When one studies the dynamics of the Yang-Mills field in such a curved space time, then apart from the gravitational connection of the base manifold, there is the Yang-Mills connection on the fibre space, the fibre space being simply identified with the Lie algebra of the Yang-Mills gauge group. The problem of determining the index of such a Dirac operator is fundamental in quantum field theory especially in the study of anomalies. The culminating point of this study is the celebrated Atiyah-Singer index theorem which expresses the index of this operator in terms of an integral of a quantity involving both the Riemann curvature tensor of the base manifold and the curvature tensor of the Yang-Mills field. This is the subject of the last chapter. Here, we also introduce the notion of Bosonic and Fermionic strings in the quantum theory based on Boson and Fermion Fock space an describe some properties of the Virasoro Lie algebra generated by the Fourier components of the energy momentum tensor of the string. Using this Virasoro algebra, we give some convincing arguments of why 26 must be the space-time dimension of the Bosonic string and ten the space-time dimension of the Fermionic string. We construct superstring actions with additional gravitino string fields in a curved two dimensional space-time that possess local supersymmetry. This action is a preliminary to our understanding of supergravity theories. Apart from that, we describe local supersymmetric actions for point

particles as a prelude to our understanding of supersymmetric string field theories. Some discussion of what a supersymmetric theory of fluid dynamics should look like is also provided.

[1] Notion of a principal bundle and a connection on a principal bundle. A Lie group G is given and a vector bundle P is given. Let $\pi : P \to M$ denote the projection of the fibre onto the base manifold of the bundle. Thus, if $\{U_i\}$ forms a chart for M, then $P|_{U_i}$ is isomorphic to $U_i \times G$ for each i ie, locally, the fibre space of the bundle is the group G itself. Another way to restructure this construction is to consider a point (x, s) in the vector bundle P where $x \in M$ is the base point and s is an element of the fibre at x. We assume that the group G acts to the right of the points of the base manifold M and to the left of the fibre. Then, we identify (x, s) with the set of all points $(xg^{-1}, gs), g \in G$. In other words, the principal bundle is locally given by $(U_i \times F)/G$ where F is the fibre space and where G acts on $U_i \times F$ as

$$g.(x, s) = (xg^{-1}, gs), g \in G$$

An example of a trivial principal G-bundle is $P = M \times V$ where M is a differentiable manifold, V is a vector space and ρ a representation of G in V. Then, G acts on the fibre space V as $v \to \rho(g)v$. The action of G on $(x, v) \in P$ is given by $g.(x, v) = (x, \rho(g)v)$. Suppose now that G also acts on M to its right. Then, the orbit of (x, v) under G is given by $(x.g^{-1}, \rho(g)v), g \in G$ and if we identify all points in an orbit, then then each G-orbit determines an element of the quotient bundle P/G. Suppose now that $M = G$. Then, $P = G \times V$ and an element of P is (g, v) with $g \in G, v \in V$. The corresponding G-orbit is $(gh^{-1}, \rho(h)v), h \in G$ and it is easy to see that this orbit can be identified with the element $(g, \rho(g)v)$ of P since $\rho(gh^{-1})\rho(h)v = \rho(g)v$. In other words, given the G-orbit in P, the elements $g \in G, v \in V$ in the pair $(g, \rho(g)v)$ are completely specified and likewise, given g, v, the orbit $(gh^{-1}, \rho(h)v), h \in G$ is completely specified.

Remarks: G is called the structure group of principal bundle P. P/G is a vector bundle. Suppose H is another Lie group and $f : G \to H$ is a homomorphism. We can then define a principal H-bundle by $P/G \times H$ where now H is the fibre and H on this fibre to its right so that it commutes with the G action via the homomorphism f acting to the left of H. This left G-action on the fibre H obviously commutes with the right H-action and hence this left G-action preserves the group structure of the fibre. As an example, consider the H-bundle $\pi : G \to G/H$ where π is the canonical projection and H is a subgroup of G. The H-orbit a point (x, y) in $G \times G$ is given by $(xh^{-1}, hy), h \in H$ and can be identified with (xH, y). In other words, the G-bundle $(G \times G)/H$ has been identified with the trivial G-bundle $G/H \times G \to G/H$. This means that the structure group H of the bundle $G \to G/H$ on extension to the structure group G makes it into a trivial G-bundle.

Connections on a vector bundle: Let P be a vector bundle and locally, let P be $U \times F$ where U is an open subset of the base manifold M and F is a vector space called the fibre space. Thus, if the bundle projection is defined by $\pi : P \to M$, then $F = \pi^{-1}(\{x\})$. Let ∇ be a connection on P. Then locally on $U \times F$, if X is a vector field having horizontal components X^i and vertical components X^a and we denote local horizontal coordinates by x^i and local vertical coordinates by y^a, we can write

$$X(x,y) = X^i(x,y)\frac{\partial}{\partial x^i} + X^a(x,y)\frac{\partial}{\partial y^a}$$

$$= X_h + X_v$$

where X_h stands for the horizontal component of X while X_v stands for its vertical component. The connection ∇ can then be expressed by its action on two such vector fields X and Y,

$$\nabla_X Y = \nabla_{X_h+X_v}(Y_h + Y_v)$$

$$= \nabla_{X_h} Y_h + \nabla_{X_h} Y_v + \nabla_{X_v} Y_h + \nabla_{X_v} Y_v$$

An example of a connection on a vector bundle: Let $M = \mathbb{R}^4$, the four dimensional space-time manifold be the base space of the bundle, and let the local fibre space be $V = \mathfrak{su}(N)$, the Lie algebra of $n \times n$ complex unitary matrices. Then if $\tau_a, a = 1, 2, ..., n$ are Hermitian generators of $\mathfrak{su}(N)$, we can define a gauge potential $A^a_\mu(x)\tau_a$ taking values in $\mathfrak{su}(N)$. Locally, the vector bundle is then $M \times V = \mathbb{R}^4 \times \mathfrak{su}(N)$ and we can define a connection on this vector bundle as

$$\gamma^b e^\mu_b(x)(\partial_\mu + iA^a_\mu(x)\tau_a) = \gamma^\mu(x)\nabla_\mu$$

where $e^\mu_b(x), b = 0, 1, 2, 3$ are vector fields (called a tetrad in general relativity) on M. This connection is called the Dirac-Yang-Mills connection. We can also include in this a spinor connection of the gravitational field $\Gamma_\mu(x)$ by defining the connection as

$$\gamma^b e^\mu_b(x)(\partial_\mu + \Gamma_\mu(x) + iA^a_\mu(x)\tau_a) = \gamma^\mu(x)\nabla_\mu$$

In this way the connection takes care of both the Riemannian curvature of the base manfiold and the Yang-Mills curvature of the fibre manifold.

Study projects: Read about

[a] Cartan's equations of structure involving expression of the torsion and curvature forms in coordinate free ways in terms of the connection form.

The Maurer-Cartan form:

De-Rham cohomology: $d : \Omega^k(M) \to \Omega^{k+1}(M), k = 0, 1, 2, ...$ is a map that takes k-differential forms on a manifold to $k+1$-differential forms and if ω is a k-form, then $d\omega$ is a k+1-form. If $d\omega = 0$, then ω is said to be a closed k-form. $d^2 = 0$ implies that exact forms are closed but there may be closed forms that are not exact. By a closed k-form ω, we mean that $d\omega = 0$ and by an exact k-form ω, we mean the existence of a $k-1$-form θ such that $\omega = d\theta$. The quotient

$H^k(\Omega, M)$ of closed k-forms by the exact k-forms is called the kth-cohomology class for the De-Rham complex.

[1] Let H be a Hamiltonian operator acting in $L^2(\mathbb{R}^3)^{\otimes N}$. Consider the anti-symmetric wave function of the positions and z-spin components of N electrons given by

$$|\psi>=\sum_{\sigma\in S_N} sgn(\sigma)|\psi_{\sigma 1}\otimes...\otimes\psi_{\sigma N}>\otimes|\eta>$$

where

$$\psi_k \in L^2(\mathbb{R}^3), k=1,2,...,N,$$

$$|\eta>=\sum_{\sigma\in S_N}|s_{\sigma 1}\otimes...\otimes s_{\sigma N}>$$

where

$$s_k \in \{\pm 1/2\}$$

Evaluate the variational derivative

$$\delta<\psi|H|\psi>/\delta\psi_k(\mathbf{r})^*$$

and hence formulate the time dependent Hartree-Fock equations for the N-electron atom:

$$i\partial\psi_k(t,r)/\partial t=\delta<\psi|H|\psi>/\delta\psi_k^*(t,r)$$

where now the marginal wave functions $\psi_k, k=1,2,...,N$ are assumed to be time dependent.

[2] Let M be a differentiable manifold and let a Lie group G act on it. Let $\mathfrak{g}$ denote the Lie algebra of G and for each $\xi \in \mathfrak{G}$, define the vector field $\xi_M(x), x \in M$ on M by

$$\xi_M(x)=\frac{d}{dt}exp(t\xi).x|_{t=0}$$

Now for any vector $v \in TM_x$, let $\gamma(v) \in \mathfrak{g}$ be defined by the relation

$$\gamma(v)_M(x)=v$$

γ is then evidently a linear function on TM_x for each $x \in M$ and γ is called the Cartan-Maurer form associated with (M,G). Let ρ be a representation of G in a vector space V. We denote the corresponding representation of $\mathfrak{g}$ also by ρ. Then, letting $\nabla_v = v+\rho(\gamma(v))$, it is easy to see that ∇ is a connection on $M \times V$. More precisely, if $s: M \to V$, ie, $s(x) \in V \forall x \in M$, then for any $v \in TM$, $\nabla_v s$ is again a mapping from $M \to V$ and if $f: M \to \mathbb{R}$, then

$$\nabla_v(fs)=v(f)s+f\nabla_v s$$

Now let v be a vector field on M. The action of G on M induces an action on vector fields on M, transforming the vector field v to $dL_g ovoL_g^{-1}$ or equivalently, the vector $v(x) \in TM_x$ to the vector $dL_g.v(g^{-1}x) \in TM_x$, or equivalently the vector $v(x) \in TM_x$ to the vector $dL_g v(x) \in TM_{gx}$. We assume that the vector field v is given by $v(x) = \xi_M(x), x \in M$ where $\xi \in \mathfrak{g}$. Then $\gamma(v) = \xi$. Now,

$$dL_g v(x) = dL_g \xi_M(x) = \frac{d}{dt} g.exp(t\xi).x|_{t=0}$$

$$= \frac{d}{dt} exp(t.Ad(g)\xi).g.x|_{t=0} = (Ad(g)\xi)_M(gx)$$

and hence

$$(dL_g v)(x) = dL_g v(g^{-1}x) = (Ad(g)\xi)_M(x)$$

or equivalently, in terms of vector fields,

$$dL_g v = (Ad(g)\xi)_M$$

Then, it follows that

$$\gamma(dL_g v) = Ad(g)\xi \in \mathfrak{g}, g \in G$$

From this, we deduce that if ρ is a representation of G or equivalently of $\mathfrak{g}$, then

$$\rho(\gamma(dL_g v)) = \rho(Ad(g)\xi) = Ad(\rho(g)).\rho(\xi) = \rho(g).\rho(\xi).\rho(g)^{-1}$$

Consider the function

$$g \to \rho(g)^{-1}x = f(g)$$

for some fixed $x \in V$ This function can be regarded as a function on the fibres of the principal bundle. It satisfies the equivariance property

$$f(gh) = \rho(h)^{-1}f(g), g, h \in G$$

Let v be a vertical vector, ie, $v \in \mathfrak{g}$ and consider

$$v(f)(g) + \rho(\gamma(v))f(g)$$

We assume that the representation ρ of G is defined by its action on the space of equivariant functions on G with values in V by the rule

$$(\rho(h)f)(g) = f(gh) = \rho(h)^{-1}f(g)$$

Since v is a vertical vector field, we can write $v(g) = g\xi$ for $\xi \in \mathfrak{g}$. Then, $v = \xi_G$ and we have $\gamma(v) = \xi$ and further, for $f(g) = \rho(g)^{-1}x$, we have

$$v(f)(g) = \frac{d}{dt} f(g.exp(t\xi))|_{t=0} = \frac{d}{dt} \rho(g.exp(t\xi))^{-1}x|_{t=0}$$

$$= -\rho(\xi)\rho(g)^{-1}x = -\rho(\xi)f(g)$$

On the other hand,

$$\rho(\gamma(v))f(g) = \rho(\xi)f(g)$$

and hence we obtain the remarkable result:

$$(v + \rho(\gamma(v)))f(g) = 0$$

In other words, the covariant derivative $v+\rho(\gamma(v))$ vanishes when v is a vertical vector field on a principal bundle.

Reference: S.Ramanan, Global Calculus, American Mathematical Society.

[3] The Electroweak theory explains how an electron acquires mass, how the gauge Bosons that propagate the nuclear forces acquire mass and it also simultaneously explains why the photon is massless, ie, does not acquire any mass after symmetry breaking occurs when one considers vacuum expectations of the Higgs doublet interacting via a Yukawa coupling to the electrons and leptons. These explanations are based on the gauge group $SU(2)_L \times U(1)$ and its symmetry breaking for the electron type leptons, the electrons and the gauge Bosons. Further, it explains why when one includes the three quarks into the interaction picture, the gauge group gets enlarged to $SU(3) \times SU(2) \times U(1)$ and is known as the electro-weak-strong unification. To discuss all these matters, we first write down the Lagrangian density of the electrons and electron type neutrino totally forming the lepton field. We denote by ν_e, the electron type neutrino field and by e the electron field. The left handed neutrino field is $\nu_{eL} = (1+\gamma_5)\nu_e/2$ and we note that

$$e_L = (1+\gamma_5)e/2, e_R = (1-\gamma_5)e/2$$

The total lepton field is

$$l_e = \begin{pmatrix} \nu_e \\ e \end{pmatrix}$$

and its Lagrangian interacting with the Bosonic gauge fields is

$$L_1 = \bar{l}_e \gamma^\mu (i\partial_\mu + t^a A^a_\mu + yB_\mu) l_e$$

where

$$(t^a)_{a=1}^3 = g((1+\gamma_5)/2)(\sigma_1, \sigma_2, \sigma_3)$$

$$y = g'(c_1((1+\gamma_5))\sigma_3 + c_2(1-\gamma_5))$$

$$t_{3R} = c_3(1-\gamma_5)$$

where c_1, c_2, c_3 are some real constants. It should be noted that the space of 4×4 matrices spanned by $t^a, a = 1,2,3, y, t_{3R}$ is the same as the space of matrices spanned by $t^a, a = 1,2,3, t_{3L}, t_{3R}$ where

$$t_{3L} = c_3(1+\gamma_5)I_2$$

The corresponding gauge group Lie algebra is thus

$$SU(2)_L \times U(1)_L \times U(1)_R$$

So we can define another Lie algebra element of this gauge group by

$$n_e = c_4(1+\gamma_5)\sigma_3 + c_6(1-\gamma_5)$$

and consider the gauge Lie algebra as being spanned by the basis elements $t^a, a = 1, 2, 3, t, n_e$. Corresponding to this Lie algebra, there should therefore be five gauge fields and not four. However, the presence of a fifth gauge field that couples to n_e has not been detected in the laboratory and so we have to work only with four gauge fields corresponding to the Lie algebra elements $t^a, a = 1, 2, 3, y$. There is a particular linear combination of y and t_3 of the form

$$q = et_3/g - ey/g' = -e\begin{pmatrix} 0 & \mathbf{0}^T \\ \mathbf{0} & I_3 \end{pmatrix}$$

which corresponds to the electronic charge. This portion of the symmetry is left unbroken even after a Higgs coupling followed by symmetry breaking when one takes the vacuum expectations of the Higgs field. Denoting

$$c = cos(\theta), s = sin(\theta)$$

we define the following transformed gauge fields

$$W_\mu = (A^1_\mu + iA^2_\mu)/\sqrt{2}, W^*_\mu = (A^1_\mu - iA^2_\mu)/\sqrt{2}$$

$$A_\mu = c.A^3_\mu - s.B_\mu, Z_\mu = s.A_\mu + c.B_\mu$$

or equivalently,

$$A^3_\mu = c.A_\mu + s.Z_\mu, B_\mu = -s.A_\mu + c.Z_\mu,$$

$$A^1_\mu = (W_\mu + W^*_\mu)/\sqrt{2}, A^2_\mu = (W_\mu - W^*_\mu)/i\sqrt{2}$$

and then we find that

$$t^a A^a_\mu + yB_\mu =$$

$$t^1(W_\mu + W^*_\mu)/\sqrt{2} + t^2(W_\mu - W^*_\mu)/i\sqrt{2}$$

$$+t^3(cA_\mu + s.Z_\mu) + y(-sA_\mu + c.Z_\mu)$$

$$= ((t^1 - it^2)/\sqrt{2})W_\mu + ((t^1 + it^2)/\sqrt{2})W^*_\mu$$

$$+(ct^3 - sy)A_\mu + (st^3 + cy)Z_\mu$$

and we see that if the Weinberg angle θ is chosen appropriately in terms of the coupling constants g, g', then the coefficient of A_μ is

$$ct^3 - sy = q$$

This suggests to us that the gauge field A_μ should be the electromagnetic/photon field carrying a charge q. q

We assume the presence of four real gauge fields $A^a_\mu, a = 1, 2, 3, B_\mu$. The total gauge Lagrangian is

$$L_2 = (-1/4)(A^a_{\nu,\mu} - A^a_{\mu,\nu} + C(abd)A^b_\mu A^d_\nu)^2 - (1/4)(B_{\nu,\mu} - B_{\mu,\nu})^2$$

Next, the Higgs doublet field

$$\phi = \begin{pmatrix} \phi_1 \\ \phi_2 \end{pmatrix}$$

contributes the following Lagrangian after taking into account its interaction with the gauge Bosons:

$$L_3 = (1/2)((i\partial_\mu + t^a A^a_\mu + yB_\mu)\phi)^*(i\partial^\mu + t^a A^{\mu a} + yB^\mu)\phi)$$

$$= (1/2)|(i\partial_\mu + t^a A_{\mu a} + yB_\mu)\phi|^2$$

The matrix $t^a A^a_\mu + yB_\mu$ can be represented as a linear combination of the gauge fields $W_\mu, W^*_\mu, Z_{,\mu}, A_\mu$ and it is clear that when the Higgs doublet ϕ falls to a vacuum state with expectation value $< \phi >$, then L_3 gives rise to a term $|(t^a A^a_\mu + yB_\mu) < \phi > |^2$ which can be expressed as a real valued quadratic combination of the fields $W_\mu, W^*_\mu, Z_\mu, A_\mu$ and it can be shown that when the vacuum expectation value $< \phi >$ of the Higgs doublet is appropriately chosen, then the coefficient of $W^*_\mu W^\mu$ and $Z_\mu Z^\mu|$ are non-zero while all the other quadratic terms including $A_\mu A^\mu$ have zero coefficients. This in particular implies that the W and Z-Bosons that communicate the nuclear forces are massive while the photon A_μ that communicates the electromagnetic forces is non-massive. Further, the expression

$$L_1 = \bar{l}_e \gamma^\mu (i\partial_\mu + t^a A^a_\mu + yB_\mu) l_e$$

for the electron-lepton Lagrangian shows that the electron has no mass. So there must also be some mechanism of symmetry breaking by which the electron acquires mass. Such a term is given by

$$L_4 = g_0(\bar{l}_e)_L \phi . e_R = g_0 \begin{pmatrix} \nu_e \\ e \end{pmatrix}^* ((1+\gamma^5)/2)\gamma^0(\phi \otimes e)$$

Note that ν_e is 4×1, e is also 4×1 and ϕ is 2×1. Thus, l_e is 8×1 and $\phi \otimes e$ is also 8×1 and hence L_4 is a scalar as it should be. When the Higgs doublet ϕ falls to the ground state with vacuum expected value $< \phi >$, then L_4 contributes a term that is quadratic in the electron field e and it is precisely this mechanism that accounts for the electronic mass. This completes the story of the electroweak theory based on the gauge group $SU(2) \times U(1)$. If, in addition, we are to describe quarks out of which the protons and neutrons in the nuclei are built, we must introduce the gauge group $SU(3)$ corresponding to the three flavours of quarks that are observed in nature. This describes the so called strong interactions along with the electro-weak interactions and the appropriate gauge group for doing this is $SU(3) \times SU(2) \times U(1)$. The coupling constants for the Lie algebra of this group must be derived from a fundamental coupling constant. The strong interactions are described by means of a Hadronic current that interacts with the current in the electro-weak theory and these interactions break the symmetry leading to the quark masses. Specifically, if Q denotes a three component quark, (each having four spinor components), then the associated quark current is given by

$$J_Q^\mu = \bar{Q}\gamma^\mu Q = Q^*\gamma^0\gamma^\mu Q$$

There may be left handed quarks, right handed quarks or a combination of both. For example in the case of left handed quarks, we would use the current

$$J_{QL}^\mu = Q_L^*\gamma^0\gamma^\mu Q_L = Q^*((1+\gamma^5)/2)\gamma^0\gamma^\mu Q$$
$$= Q^*((1+\gamma^5)/2)\gamma^0\gamma^\mu((1+\gamma^5)/2)Q$$
$$= Q^*\gamma^0\gamma^\mu((1+\gamma^5)/2)Q$$

since γ^5 anticommutes with γ^μ and hence commutes with $\gamma^0\gamma^\mu$. Likewise, for right handed quarks, the appropriate current would be

$$J_{QR}^\mu = Q_R^*\gamma^0\gamma^\mu Q_R = Q^*((1-\gamma^5)/2)\gamma^0\gamma^\mu Q$$

The leptonic current of the electro-weak theory on the other hand is

$$J_{l_e}^\mu = l_e^*\gamma^0\gamma^\mu l_e$$

and the interaction between these two currents that breaks the symmetry of the gauge group $SU(3)\times SU(2)\times U(1)$ is given by

$$J_Q^\mu J_{\mu l_e}$$

It should be noted that in the absence of such hadron-Lepton currents that break the gauge symmetry, we have to embed both the electron like Leptons and the quarks into this large gauge group. All this leads us to the standard model for non-supersymmetric particle physics namely the "Electroweak-Strong Unification".

We refer to Steven Weinberg's book, "The quantum theory of fields vol.II" for further details.

[4] Show that the general solution to string equations both in the Bosonic and Fermionic cases can be expressed as the superposition of forward and backward propagating strings, the forward propagating string being a function of $\tau-\sigma$ and the backward propagating string being a function of $\tau+\sigma$. Assuming the canonical equal time commutation relations for the Bosonic string and the canonical anticommutation relations for the Fermionic string, derive the corresponding commutation relations for the Bosonic creation and annihilation operators that appear in the Fourier expansions of the Bosonic string and the corresponding anticommutation relations for the creation and annihilaition operators appearing in the Fourier expansions of the Fermionic string.

[5] Consider the vertex operator for the Bosonic string

$$V(k,z) =: exp(ik.X(z)) := exp(ik.\sum_{n<0}\alpha(n)z^n/n).exp(ik.\sum_{n>0}\alpha(n)z^n/n)$$

Note that $\alpha^\mu(n), n > 0$ are the Boson annihilation operators while $\alpha(n), n < 0$ are the creation operators. They satisfy the CCR

$$[\alpha^\mu(n), \alpha^\nu(m)] = \eta^{\mu\nu} n\delta[n+m]$$

Let $|w>$ be a coherent state, ie,

$$\alpha^\mu(n)|w>= w^\mu(n)|w>, ((w^\mu(n)))_{n>0} = w^\mu, w^\mu(n) \in \mathbb{C}$$

Then,

$$V(k,z)|w>= exp(ik.\sum_{n<0} \alpha(n)z^n/n).exp(ik.\sum_{n>0} w(n)z^n/n)$$

and hence if $|u>$ is another coherent state, then using $\alpha(n)^* = \alpha(n)$,

$$<u|V(k,z)|w>= exp(ik\sum_{n<0} u(-n)\bar{z}^n/n).exp(ik.\sum_{n>0} w(n)z^n/n)$$

[6] Show that if A, B are operators in a Hiibert space such that $C = [A, B]$ commutes with both A and B, then

$$F(t) = exp(t(A+B)) = exp(tA)G(t)$$

$$exp(tA)(G'(t) + AG(t)) = (A+B)exp(tA)G(t)$$

so that

$$G'(t) = exp(-t.ad(A))(B)G(t) = (B - t[A,B])G(t)$$

and hence

$$G(t) = exp(-t^2[A,B]/2).exp(tB)$$

so that

$$exp(A+B) = exp(-[A,B]/2).exp(A).exp(B)$$

Interchanging A and B in this expression gives

$$exp(A+B) = exp([A,B]/2)exp(B).exp(A)$$

Thus, we obtain the Weyl commutation relations

$$exp(A).exp(B) = exp([A,B])exp(B).exp(A)$$

Now for a Bosonic string, take

$$A = k.\sum_{n>0} \alpha(n)z^n/n, B = k.\sum_{n<0} \alpha(n)z^n/n$$

Then,

$$[A,B] = k^2\sum_{n>0} |z|^{2n}/n, k^2 = k.k = \eta_{\mu\nu}k^\mu k^\nu$$

and hence conclude that

$$exp(k.X(z)) = exp((k^2/2). : exp(k.X(z)) :$$

Remark: $: F :$ means "normal ordering of F, ie, assuming that F is a function of the creation and annihilation operators, $: F :$ is obtained from F by replacing each term in the Taylor expansion of F with the corresponding term in which all the annihilation operators have been pushed to the right and all the creation operators to the left.

[7]

$$< 0|X^\mu(z)X^\nu(w)|0 >=< 0|(X_+^\mu(z) + X_-^\mu(z))(X_+^\nu(w) + X_-^\nu(w))|0 >=$$

$$< 0|X_+^\mu(z)X_-^\nu(w)|0 >=< 0[X_+^\mu(z), X_-^\nu(w)]|0 >$$

since

$$X_+^\mu(z)|0 >= 0, < 0|X_-^\mu(z) = 0$$

Further,

$$[X_+^\mu(z), X_-^\nu(w)] = [\sum_{n>0} \alpha^\mu(n)z^n/n, \sum_{n<0} \alpha^\nu(n)w^n/n]$$

$$= \eta^{\mu\nu} \sum_{n>0} (zw)^n/n = -\eta^{\mu\nu} log(1 - zw)$$

Note that this is not the string propagator. The string propagator is

$$D^{\mu\nu}(z, w) =< 0|T(X^\mu(z)X^\nu(w))|0 >$$

where $z = exp(\tau + \sigma), w = exp(\tau' + \sigma')$ and the T stands for the Bosonic time ordering, ie,

$$T(X^\mu(z).X^\nu(w)) = \theta(\tau - \tau')X^\mu(z)X^\nu(w) + \theta(\tau' - \tau)X^\nu(w)X^\mu(z)$$

From the string differential equations

$$(\partial_\tau^2 - \partial_\sigma^2)X(z) = 0$$

it is easy to see just as in four dimensional quantum field theory that the string propagator satisfies the differential equation

$$(\partial_\tau^2 - \partial_\sigma^2)D^{\mu\nu}(z, w) = \eta^{\mu\nu}\delta(\tau - \tau')\delta(\sigma - \sigma')$$

which has the solution

$$D^{\mu\nu}(z, w) = K.log(\sqrt{(\tau - \tau')^2 + (\sigma - \sigma')^2})$$

Note: Replace σ, σ' by $i\sigma, i\sigma'$ respectively and then the above Green's function equation becomes the Green's function equation for the two dimensional Laplacian operator whose solution is well known or equivalently can be derived from the Green's function equation for the three dimensional Laplacian by integrating w.r.t. the third coordinate.

[8] We now consider a Bosonic string having in addition to its Fock space degrees of freedom, the Hilbert space in which the position x of the centre of mass of the string as well as the momentum p of its centre of mass are taken into account. At time τ and at the spatial location σ, the string displacement field is given by

$$X^\mu(\tau,\sigma) = x^\mu + p^\mu(\tau-\sigma) - i\sum_{n\neq 0}\frac{\alpha^\mu(n)}{n}exp(in(\tau-\sigma))$$

$$= x^\mu - ip^\mu.ln(z) + i\sum_{n\neq 0}\frac{\alpha^\mu(n)}{n}z^n$$

where

$$z = exp(i(\tau-\sigma))$$

Note that this is a Hermitian operator valued string field since

$$\alpha^\mu(n)^* = \alpha^\mu(-n)$$

Then we find that with $|0>$ denoting the vacuum state of the Boson Fock space tensored with the zero eigenvalue state of the annihilation operator of the centre of mass ie of the operator $a^\mu = x^\mu + ip^\mu$,

$$[x^\mu, p^\nu] = i\eta^{\mu\nu}$$

so that

$$[a^\mu, a^{\nu *}] = -2\eta^{\mu\nu}$$

$$x^\mu - ip^\mu.ln(z) = (a^\mu + a^{\mu *})/2 - (a^\mu - a^{\mu *})ln(z)/2$$

$$= (1/2)a^\mu(1-ln(z)) + (1/2)a^{\mu *}(1+ln(z))$$

so that

$$< 0|X^\mu(\tau,\sigma).X^\nu(\tau',\sigma')|0 >=$$

$$[(1/2)a^\mu(1-ln(z)),(1/2)a^{\nu *}(1+ln(z'))] + [X_+^\mu(z), X_-^\nu(z')]$$

$$= -(1/2)(1-(ln(z))(1+ln(z'))\eta^{\mu\nu} - \sum_{n>0} z^n z'^{-n}/n$$

$$-(1/2)(1-(ln(z))(1+ln(z'))\eta^{\mu\nu} + ln(1-z/z')$$

[9] The energy-momentum tensor of a superstring. The Lagrangian is

$$L = (1/2)\partial_\alpha X^\mu \partial^\alpha X_\mu + i\psi_+\partial_-\psi_+$$

$$+i\psi_-\partial_+\psi_-$$

where

$$\sigma^0 = \tau, \sigma^1 = \sigma, \sigma^+ = \tau + \sigma, \sigma^- = \tau - \sigma,$$

$$\partial_+ = \partial_\tau + \partial_\sigma, \partial_- = \partial_\tau - \partial_\sigma$$

ψ_+ is assumed to be a function of only σ^+ and ψ_- is a function of only σ^-. Thus, we have the obvious Fermionic field equations

$$\partial_- \psi_+ = 0, \partial_+ \psi_- = 0$$

which may also be derived from the Lagrangian. Now the components of the energy-momentum tensor are in the $\tau - \sigma$ coordinate system given by

$$T_0^0 = (\partial L / \partial \psi_{+,\tau}) \psi_{+,\tau} + \partial L / \partial \psi_{-,\tau} \psi_{-,\tau} - L$$

$$= i\psi_+ \psi_{+,\tau} + i\psi_- \psi_{-,\tau} - i(\psi_+ \partial_- \psi_+ + \psi_- \partial_+ \psi_-)$$

$$= i\psi_+ \psi_{+,+} + i\psi_- \psi_{-,-} - i(\psi_+ \partial_- \psi_+ + \psi_- \partial_+ \psi_-)$$

$$= i(\psi_+ \partial_+ \psi_+ + \psi_- \partial_- \psi_-)$$

provided that we assume that the field equations are satisfied. The total energy in the Fermionic component of the superstring is then

$$H_F = \int_0^{2\pi} T_0^0 d\sigma$$

The solution to the Fermionic field equations are

$$\psi_+(\sigma^+) = \sum_n S_n exp(in\sigma^+), \psi_-(\sigma^-) = \sum_n \tilde{S}_n.exp(in\sigma^-)$$

The canonical anticommutation relations are obtained as

$$\pi_+ = \partial L / \partial \psi_{+,\tau} = \partial L / \partial \psi_{+,-} = i\psi_+$$

$$\pi_- = \partial L / \partial \psi_{-,\tau} = \partial L / \partial \psi_{-,+} = i\psi_-$$

and hence using

$$\{\psi_+(\tau + \sigma), \pi_+(\tau + \sigma')\} = i\delta(\sigma - \sigma')$$

and

$$\{\psi_-(\tau + \sigma), \pi_-(\tau + \sigma')\} = i\delta(\sigma - \sigma')$$

we get

$$\{\psi_+(\sigma^+), \psi_+(\sigma'^+\} = \delta(\sigma^+ - \sigma'^+)$$

$$\{\psi_-(\sigma^-), \psi_-(\sigma'^-\} = \delta(\sigma^- - \sigma'^-)$$

These imply the following anticommutation relations for $\{S_n\}$ and $\{\tilde{S}_n\}$:

[10] Virasoro algebra and the critical dimension D=26 for the Bosonic string.

$$[\alpha^\mu(m), \alpha^\nu(n)] = \eta^{\mu\nu}.m.\delta(m + n)$$

This follows from the CCR for the Bosonic string:

$$X^\mu(z) = -i\sum_{n\neq 0}(\alpha^\mu(n)/n)z^n, z = exp(in(\tau-\sigma)),$$

$$[X^\mu(z), \partial_\tau X^\nu(z')] = i\eta^{\mu\nu}\delta(\sigma-\sigma'), z = \tau-\sigma, z' - \tau - \sigma'$$

The energy momentum tensor of the string is

$$T^\alpha_\beta = (\partial L/\partial X^\mu_{,\alpha})X^\mu_{,\beta} - L\delta^\alpha_\beta$$

$$= \partial^\alpha X_\mu.\partial_\beta X_\mu - \delta^\alpha_\beta L$$

$$L = (1/2)\partial^\alpha X_\mu.\partial_\alpha X_\mu$$

We have with

$$\sigma^+ = \tau+\sigma, \sigma^- = \tau-\sigma,$$

so that

$$\sigma^0 = \tau = (\sigma^+ + \sigma^-)/2, \sigma^1 = \sigma = (\sigma^+ - \sigma^-)/2,$$

$$T^+_+ = (1/2)(T^0_0 + T^1_1 - T^1_0 - -T^0_1),$$

$$T^-_- = (1/2)(T^0_0 + T^1_1 - T^1_0 - T^0_1) = T^+_+,$$

$$T^+_- = (1/2)(T^0_0 - T^1_1 + T^1_0 - T^0_1)$$

$$T^-_+ = (1/2)(T^0_0 - T^1_1 - T^1_0 + T^0_1)$$

These formulas follow directly from the rules of tensor transformation under a change of the coordinate system. An easier way to compute these is to express the Lagrangian as

$$L = (1/2)\partial_+ X^\mu.\partial_- X_\mu$$

and then observe that

$$T^+_+ = (\partial L/\partial\partial_+ X^\mu)(\partial_+ X^\mu) - L$$

$$= (1/2)\partial_+ X^\mu.\partial_- X_\mu$$

$$T^-_- = (\partial L/\partial\partial_- X^\mu).(\partial_- X^\mu) - L$$

$$= (1/2)\partial_+ X_\mu.\partial_- X^\mu = T^+_+$$

$$T^+_- = (\partial L/\partial\partial_+ X^\mu)(\partial_- X^\mu)$$

$$= (1/2)\partial_- X_\mu.\partial_- X^\mu$$

$$T^-_+ = (\partial L/\partial\partial_- X^\mu)(\partial_+ X^\mu) =$$

$$(1/2)\partial_+ X_\mu.\partial_+ X^\mu$$

Now writing

$$X^\mu(\sigma^+, \sigma^-) = (-i)\sum_{n\neq 0}(\alpha^\mu(n)/n)exp(in\sigma^-)$$

$$+(-i)\sum_{n\neq 0}\beta^{\mu}(n)/n)exp(in\sigma^{+})$$

we get

$$\partial_{+}X^{\mu}=\sum_{n}\beta^{\mu}(n)exp(in\sigma^{+}),$$

$$\partial_{-}X^{\mu}=\sum_{n}\alpha^{\mu}(n)exp(in\sigma^{-})$$

We then find using the CCR

$$[X^{\mu}(\tau,\sigma),\partial_{\tau}(X^{\nu}(\tau,\sigma')]=i\delta(\sigma-\sigma')\eta^{\mu\nu}$$

that

$$[\alpha^{\mu}(n),\alpha^{\nu}(m)]=\eta^{\mu\nu}\delta(n+m),$$

$$[\beta^{\mu}(n),\beta^{\nu}(m)]=\eta^{\mu\nu}\delta(n+m),$$

$$[\alpha^{\mu}(n),\beta^{\nu}(m)]=0$$

and

$$T_{+}^{+}=(1/2)\sum_{n,m}\alpha^{\mu}(n)\beta_{\mu}(m)exp(i(n(\tau+\sigma)+m(\tau-\sigma))$$

$$=(1/2)\sum_{n,m}\alpha^{\mu}(n)\beta_{\mu}(m)exp(i(n+m)\tau+i(n-m)\sigma)$$

$$=T_{-}^{-}$$

$$T_{-}^{+}=(1/2)\sum_{n,m}\alpha^{\mu}(n).\alpha_{\mu}(m)exp(i(n+m)\sigma^{-})$$

$$=(1/2)\sum_{n,m}\alpha^{\mu}(n-m)\alpha_{\mu}(m)exp(in\sigma^{-})$$

$$=\sum_{n}L_{n}exp(in\sigma^{-})$$

where

$$L_{n}=(1/2)\sum_{m}\alpha(n-m).\alpha(m)=(1/2)\sum_{m}\alpha(n+m).\alpha(-m)$$

$$T_{+}^{-}=\sum_{n}\tilde{L}_{n}.exp(in\sigma^{+})$$

where

$$\tilde{L}_{n}=(1/2)\sum_{m}\beta(n-m).\beta(m)$$

Note that the operators $L_n,\tilde{L}_n$ are well defined for $n\neq 0$. For $n=0$ however, there is a problem since in the expressions

$$L_{0}=(1/2)\sum_{n}\alpha(n).\alpha(-n),\tilde{L}_{0}=(1/2)\sum_{n}\beta(n).\beta(-n)$$

the operators $\alpha(n)$ and $\alpha(-n)$ do not commute as also with $\beta(n)$ and $\beta(-n)$. Note that

$$\alpha(0,\tau)$$

should be interpreted as the limit of $-i\alpha(n)exp(in\tau)/n$ with $n \to 0$. This gives $\alpha(0,\tau) = \alpha(0)\tau = p\tau$ and hence we have the expression

$$X^\mu(\tau,\sigma) = x^\mu + p^\mu\tau + \sum_{n\neq 0}(\alpha^\mu(n)/n)exp(in\sigma^-) + \sum_{n\neq 0}(\beta^\mu(n)/n)exp(in\sigma^+)$$

where

$$[x^\mu, p^\nu] = i\eta^{\mu\nu}$$

and x^μ, p^μ commute with $\alpha^\nu(n), \beta^\nu(n)$ for all ν and all $n \neq 0$. We define $L_0, \tilde{L}_0$ using the normal ordering formalism:

$$L_0 = p^2/2 + \sum_{n\geq 1}\alpha(-n).\alpha(n),$$

$$\tilde{L}_0 = p^2/2 + \sum_{n\geq 1}\beta(-n).\beta(n)$$

Then, we evaluate

$$4[L_n, L_m] = [\sum_k \alpha(n-k).\alpha(k), \sum_r \alpha(m-r).\alpha(r)]$$

$$= \sum_{k,r}[\alpha(n-k).\alpha(k), \alpha(m-r).\alpha(r)]$$

$$= \sum_{k,r}[\alpha^\mu(n-k), \alpha^\nu(m-r)]\alpha_\mu(k)\alpha_\nu(r)$$

$$+\sum_{k,r}\alpha^\mu(n-k)[\alpha_\mu(k), \alpha^\nu(m-r)]\alpha_\nu(r)$$

$$+\sum_{k,r}\alpha^\nu(m-r)[\alpha^\mu(n-k), \alpha_\nu(r)]\alpha_\mu(k)$$

$$+\sum_{k,r}\alpha^\mu(n-k)\alpha^\nu(m-r)[\alpha_\mu(k), \alpha_\nu(r)]$$

Evaluating these commutators using the CCR gives us

$$[L_m, L_n] = (m-n)L_{m+n} + A(m)\delta(m+n)$$

where $A(m)$ is a scalar. This scalar central charge term $A(m)$ arises because of the ambiguity in defining L_0. L_0 is defined as

$$L_0 = (1/2)\sum_n \alpha(-n).\alpha(n)$$

but could also after after commuting $\alpha(-n)$ with $\alpha(n)$ be defined with a reversed order after adding an infinite constant or even more generally, we could change the order in which $\alpha(n)$ and $\alpha(-n)$ appear for a finite or infinite set of $n's$. Thus, we cannot just write $[L_n, L_{-n}] = 2nL_0$ as we could do with $[L_n, L_m] = (n-m)L_{n+m}$ when $n+m \neq 0$. In fact when $m \neq 0$, $\alpha(n+m)$ commutes with $\alpha(-n)$ and there is no ordering ambiguity in defining

$$L_m = \sum_n \alpha(n+m).\alpha(-n)$$

Now it turns out that if we make use of the Jacobi identity for Lie brackets in the form

$$[L_m, [L_r, L_s]] + [L_r, [L_s, L_m]] + [L_s, [L_m, L_r]] = 0$$

then we get a difference equation for $A(m)$ which on solving gives us

$$A(m) = D(m^3 - m)/12$$

where D is the space-time dimension. Then, we define a physical state as that which is an eigenstate for L_0 all with a given fixed energy a(the Bosonic energy operator) and which is annihilated by $L_m, m > 0$. A spurious state is defined to be a state that is eigen to L_0 with the same energy a and is simultaneously orthogonal to all the physical states. Thus, if a state is both physical and spurious, it must have zero norm. When $D = 26$, the number of zero norm states increases dramatically and since zero norm states are preferred, the critical Bosonic dimension is 26. It should be noted that non-zero zero norm states occur in string theory because the commutation relations

$$[\alpha^\mu(n), \alpha^\nu(m)] = \eta^{\mu\nu} n\delta(n+m)$$

and

$$\alpha^\mu(-m) = \alpha^{\mu *}(m)$$

imply that the vacuum state $|0>$ is annihilated by all the $\alpha^\mu(n) > 0$, and hence satisfies

$$\alpha^r(n)\alpha^{r*}(n)|0>= -|0>, r > 0$$

and hence

$$< 0|\alpha^r(n)\alpha^{r*}(n)|0>= -1$$

which means that the underlying Boson Fock space is a not a proper Hilbert space in the correct sense. This is a consequence of using the Minkowski metric of special relativity rather than a Euclidean metric. Likewise the Fermionic space-time dimension comes out to be 10 on the grounds that the super-Yang-Mills Lagrangian has local supersymmetry only when the Dirac Gamma matrices satisfy a certain identity that is valid only in 10 space-time dimensions. Another confirmation of the ten dimensionality of space-time for Fermionic strings comes from the same argument as that used for Bosonic strings. At precisely this critical dimension, one obtains a very large number of zero norm states constructed from the Fourier series components of the Fermionic energy-momentum tensor.

[11] Some remarks related to string field theory and supersymmetric field theories in higher dimensional space-time. String field theory talks about the motion of individual classical and quantum strings having one time and one space degree of freedom. The number of components of the string in the Bosonic case comes out to be 26 when we consider a situation in which one is required to have a maximum family of zero norm physical states. Likewise in the Fermionic string case, the number of components of the string comes out to be 10. When one considers a superstring comprising of a Bosonic and a Fermionic component, then the symmetry group comes out to be $SO(10)$. This group has two irreducible spin representations acting in vector spaces each of dimension $2^4 = 16$. On Bosonization, ie transformation of the Fermionic components into Bosonic components via quadratic relations, we find that the sixteen dimensional Bosonic string propagating in the left direction couples with the ten dimensional Bosonized string coming from Bosonization of the left moving Fermionic string to give a total of $16 + 10 = 26$ dimensional left moving Bosonic strings and a ten dimensional Fermionic string coming from the vector representation of $SO(10)$. These heuristic interpretation of the dimensions of the Bosonic and Fermionic strings are justified by the above mentioned rigorous proofs based on obtaining a maximal set of zero norm states. Zero norm states are preferred because the quantum string has the least energy in such states.

Just as in mechanics, one first talks about point particles moving in four dimensional space-time so that at any proper time, the particle's configuration is specified by four space-time variables, and then by considering a very large aggregate of such point particles which in the continuum limit becomes a matter field and then one writes equations of motion for the matter field in terms of partial differential equations in four dimensional space-time, likewise, in string theory, one first considers the motion of individual strings whose configuration is specified by ten coordinates which are functions of two variables, one a proper time variable and two a length variable and then one passes to the continuum limit of a very large aggregate of such strings which are then described by field equations expressed in the form of partial differential equations in ten dimensional space-time for functions of ten variables. Thus, one obtains the super-Yang-Mills field theory as a low energy limit of superstring theory. As in the case of point particles, so in the case of strings, if we have an continuum aggregate of strings, then writing down the two variable partial differential equations of motion of each individual string becomes very cumbersome and hence one uses the low energy approximation corresponding to a complex distribution over the ten dimensional string space and thereby obtains the super-Yang-mills equations as a good approximation to continuum string theory.

[12] A supersymmetric theory for point particles.

Consider the Lagrangian

$$L(x, x', \theta, \theta') = (e(x)/2)(x^{\mu'}(t) + \theta(t)^T \epsilon \Gamma^\mu \theta'(t))^2$$

where x is a Bosonic vector and θ is a Fermionic vector. square denotes the Minkowski quadratic form. Define

$$p^\mu = x^{\mu'} + \theta^T \epsilon \Gamma^\mu \theta'$$

Consider a supersymmetric transformation

$$\delta x^\mu = c_1 \theta^T \epsilon \Gamma^\mu \delta\theta$$

where

$$\delta\theta = c_2 \Gamma^\mu k p_\mu$$

where k is an infinitesimal Fermionic parameter and c is a constant. We have

$$\delta p^\mu = \delta x^{\mu'} + \delta\theta^T \epsilon \Gamma^\mu \theta' + \theta^T \epsilon \Gamma^\mu \delta\theta'$$

$$= c_1 \theta^T \epsilon \Gamma^\mu \delta\theta' + c_1 \theta^{'T} \epsilon \Gamma^\mu \delta\theta$$

$$-c_2 p_\nu k^T \Gamma^{\nu T} \epsilon \Gamma^\mu \theta'$$

$$+c_2 \theta^T \epsilon \Gamma^\mu \Gamma^\nu k p'_\nu$$

$$= c_1 c_2 \theta^T \epsilon \Gamma^\mu \Gamma^\nu k p'_\nu + c_1 c_2 \theta^{'T} \epsilon \Gamma^\mu \Gamma^\nu k p_\nu$$

$$-c_2 p_\nu k^T \Gamma^{\nu T} \epsilon \Gamma^\mu \theta'$$

$$+c_2 \theta^T \epsilon \Gamma^\mu \Gamma^\nu k p'_\nu$$

By choosing of $c_1 = -1, c_2 = -c$, we can make the terms involving p'_ν, ie, the first and the last terms in the above sum cancel and then we get

$$\delta p^\mu =$$

$$c\theta^{'T} \epsilon \Gamma^\mu \Gamma^\nu k p_\nu$$

$$+c p_\nu k^T \Gamma^{\nu T} \epsilon \Gamma^\mu \theta'$$

$$= c\theta^{'T} \epsilon \Gamma^\mu \Gamma^\nu k p_\nu$$

$$+c p_\nu k^T \epsilon \Gamma^\nu \Gamma^\mu \theta'$$

$$= c p_\nu k^T [\Gamma^{\nu T} \Gamma^{\mu T} \epsilon + \epsilon \Gamma^\nu \Gamma^\mu] \theta'$$

$$= 2c p_\nu k^T \epsilon \Gamma^\nu \Gamma^\mu \theta'$$

where we have made use of the anticommutativity of k and θ', and the skew-symmetry of ϵ and $\epsilon\Gamma^\mu$. Thus,

$$\delta(p^2/2) = p_\mu \delta p^\mu =$$

$$2c p_\mu p_\nu k^T \epsilon \Gamma^\nu \Gamma^\mu \theta'$$

$$= c p^2 k^T \epsilon \theta'$$

It follows that the total variation in the Lagrangian is

$$\delta L = e\delta(p^2/2) + (p^2/2)\delta e$$

$$= p^2/2(\delta e + 2eck^T\epsilon\theta')$$

Therefore, if we define the supersymmetry transformation of e by

$$\delta e = -2ck^T\epsilon\theta'$$

we get

$$\delta L = 0$$

ie, supersymmetry invariance of the Lagrangian.

Remark: Suppose that that infinitesimal Fermionic parameter $k = k(t)$ depends on time. Then also we would get invariance of the action. Indeed in this case, the terms in the variation of the action that depend on k' come from

$$\delta x^{\mu'} = c_1\theta^T\epsilon\Gamma^\mu\delta\theta'$$

and from

$$\theta^T\epsilon\Gamma^\mu\delta\theta'$$

These terms are respectively

$$c_1c_2\theta^T\epsilon\Gamma^\mu\Gamma^\nu k'p_\nu$$

and

$$c_2\theta^T\epsilon\Gamma^\mu\Gamma^\nu k'p_\nu$$

and obviously, the sum of these two terms is zero. Hence, our action has not only global supersymmetry invariance but also local supersymmetry invariance.

[13] An application of superstrings:
Given a superstring

$$X^\mu(\tau,\sigma) = x^\mu + p^\mu\tau - i\sum_{n\neq 0}(\alpha^\mu(n)/n)exp(in(\tau-\sigma))$$

$$-i\sum_{n\neq 0}(\beta^\mu(n)/n)exp(in(\tau+\sigma))$$

$$\psi^\mu(\tau,\sigma) = \sum_n S_n^\mu exp(in(\tau-\sigma)) + \sum_n \tilde{S}_n^\mu exp(in(\tau+\sigma))$$

where the CCR's

$$[\alpha^\mu(n),\alpha^\nu(m)] = \eta^{\mu\nu}n\delta(n+m),$$

$$[\beta^\mu(n),\beta^\nu(m)] = \eta^{\mu\nu}n\delta(n+m),$$

$$[\alpha^\mu(n),\beta^\nu(m)] = 0,$$

and the CAR's

$$\{S_n^\mu, S_m^\nu\} = \eta^{\mu\nu}\delta(n+m)$$

we wish to add an extra control field contribution to the total Lagrangian and then control this field in order to obtain a desired quantum unitary gate. The question is what kind of additional fields with interactions with the Bosonic and Fermionic components can be added while yet retaining global/local supersymmetry invariance ?

[14] **Adding Fermionic terms to Bosonic ordinary differential equations**

Consider a one dimensional Bosonic ode

$$x'(t) = f(t, x(t)), x(t) \in \mathbb{R}^n, f : \mathbb{R}_+ \times \mathbb{R}^n \to \mathbb{R}^n$$

If $n = 2p$, this differential equation may, for example, be derived from a Hamiltonian $H(q,p)$ in p position and p momentum variables and hence from a Lagrangian as a function of p coordinates and their derivatives $L_B(t,q,q')$. We now add Fermionic terms to this Lagrangian so that the total Lagrangian becomes

$$L(t,q,q',\theta,\theta') = L_B(t,q,q') + L_F(t,q,q',\theta,\theta')$$

so that the total Lagrangian is invariant under a global infinitesimal supersymmetry transformation of the form

$$\delta q^m = \theta^T \epsilon \Gamma^m(q).\delta\theta,$$

$$\delta\theta = \Gamma^m(q)k(t)p_m$$

where $k(t)$ is an infinitesimal Fermionic parameter that depends on time and $\Gamma^m(q)$ are position dependent matrices. Here,

$$p_m = \partial L / \partial q^{m'}$$

How to choose the matrices Γ^m so that (a) the above Lagrangian changes only by a total time derivative under this supersymmetry transformation and (b) these infinitesimal supersymmetry transformations form a Lie super algebra so that we have a super Lie group associated with these transfromations. This problem was solved above for Lagrangians whose momenta are linear in the time derivative of the position operators and bilinear in the Fermionic variables and their time derivatives. Of course, since there are in all p Fermionic variables, we cannot have a term involving a product of either more than p Fermionic parameters or more than p Fermionic velocity parameters. We can thus consider a Lagrangian of the form

$$L(q,q',\theta,\theta') = \sum_{0 \le r,s \le p} A(r,s,q,q')^T(\theta^{\otimes r} \otimes \theta'^{\otimes s})$$

and demand local supersymmetry invariance. More precisely, since all the θ's and all the θ''s in each product must be distinct, or Lagrangian must have the form

$$L(q,q',\theta,theta')$$

$$= \sum_{1 \le k_1 < ... < k_r \le p, 1 \le m_1 < ... < m_s \le p} A(k_1,...,k_r,m_1,...,m_s,q,q')\theta_{k_1}...\theta_{k_r}\theta'_{m_1}...\theta'_{m_s}$$

[15] Problem: Suppose we consider the motion of a one dimensional particle in a potential $U(x)$. Its Lagrangian is

$$L_1(x,x') = x'^2/2 - U(x)$$

We replace the Bosonic variable x by

$$x = y + (1/2)\theta^T\epsilon\Gamma\theta$$

where θ is a Fermionic variable of size $p \times 1$ and $\epsilon\Gamma$ is a real skew-symmetric $p \times p$ matrix. Then the Lagrangian becomes

$$L(y,y',\theta,\theta') = L_1(y + (1/2)\theta^T\epsilon\Gamma\theta, y' + \theta'^T\epsilon\Gamma\theta)$$

This idea or constructing super-symmetric Lagrangians for point particles is similar to the construction of supersymmetric Lagrangians in Chiral field theories wherein if Φ is for example a left Chiral field and f a function, then $[f(\Phi)]_2$ is a supersymmetric Lagrangian and so is

$$[\Phi^*\Phi]_D + [f(\Phi)]_2$$

The function f generates a superpotential term which is a function of the scalar Bosonic Klein-Gordon field. We have

$$x'^2 = y'^2 + 2y'\theta'^T\epsilon\Gamma\theta$$

and

$$U(x) = U(y) + U'(y)\theta^T\epsilon\Gamma\theta/2 + ... + U^p(y)(\theta^T\epsilon\Gamma\theta)^p/2^p p!$$

It is hard to find an appropriate supersymmetry transformation that would leave L invariant. So we go back to the simplest local supersymmetry-invariant action that we considered above

$$S[x,\theta] = (1/2)\int e(x(t))(x^{\mu'} + (1/2)\theta^T\epsilon\Gamma^\mu\theta')^2 dt$$

where by $(A^\mu)^2$, we mean $A_\mu A^\mu = \eta_{\mu\nu}A^\mu A^\nu$ with $\eta_{\mu\nu}$ being the Minkowski metric. We write down the equations of motion for this action:

$$L = (1/2)e(x(t))(x^{\mu'} + \theta^T\epsilon\Gamma^\mu\theta')^2$$

$$\partial L/\partial x^{\mu'} = e(x)(x^{\mu'} + \theta^T\epsilon\Gamma^\mu\theta')$$

$$= ep^\mu$$

$$\partial L/\partial x^\mu = (1/2)e_{,\mu}(x)p^2$$

$$\partial L/\partial\theta' = e(x)(\epsilon\Gamma_\mu\theta)p_\mu$$

$$\partial L/\partial\theta = e(x)\epsilon\Gamma_\mu\theta'.p_\mu$$

and hence, our equations of motion are

$$\frac{d}{dt}(e(x)(x^{\mu'} + \theta^T \epsilon\Gamma^\mu \theta')) =$$

$$(1/2)e^{,\mu}(x)(x^{\nu'} + \theta^T \epsilon\Gamma^\nu \theta')^2$$

$$\frac{d}{dt}(e(x)(\epsilon\Gamma_\mu \theta)p_\mu) = e(x)\epsilon\Gamma_\mu \theta'.p_\mu$$

On expanding the time derivatives using the chain rule of differential calculus, we get

$$e(x)(x^{\mu''} + 2\theta^T \epsilon\Gamma^\mu \theta'' + \theta'^T \epsilon\Gamma^\mu \theta'))+$$

$$e_{,\nu}(x)x^{\nu'}(x^{\mu'} + \theta^T \epsilon\Gamma^\mu \theta')$$

$$= (1/2)e^{,\mu}(x)(x^{\nu'} + \theta^T \epsilon\Gamma^\nu \theta')^2,$$

$$e(x)(\epsilon\Gamma^\mu \theta'.p_\mu + \epsilon\Gamma^\mu \theta.p'_\mu)$$

$$+e_{,\nu}x^{\nu'}(\epsilon\Gamma_\mu \theta)p_\mu = e\epsilon\Gamma_\mu \theta' p_\mu$$

Making cancellations in the θ equation gives

$$e(x)\epsilon\Gamma^\mu \theta.p'_\mu = e_{,\nu}(x)x^{\nu'}\epsilon\Gamma^\mu \theta.p_\mu$$

[16] Construction a locally supersymmetric action for superstrings:
[a]

$$S = c_1 \int e(x)(X^\mu_{,\alpha}X^{\mu,\alpha} - i\bar{\psi}^\mu \rho^\alpha \psi_{\mu,\alpha})d^2\sigma$$

$$+c_2 \int e(x)\bar{\chi}_\alpha \rho^\beta \rho^\alpha \psi^\mu X_{\mu,\beta} d^2\sigma$$

$$+c_3 \int e(x)\bar{\psi}_\mu \psi^\mu \bar{\chi}_\alpha \rho^\beta \rho^\alpha \chi_\beta d^2\sigma$$

where the appropriate local supersymmetry transformations are

$$\delta\chi_\alpha = \nabla_\alpha \epsilon$$

$$\delta X^\mu = \bar{\epsilon}\psi^\mu, \delta\psi^\mu = -\rho^\alpha \epsilon(X^\mu_{,\alpha} - \bar{\psi}^\mu \chi_\alpha)$$

$$\delta e^a_\alpha(x) = c_4 \bar{\epsilon}(x)\rho^a \chi_\alpha(x), a, \alpha = 0, 1$$

so that

$$e = det(e^a_\alpha), \delta e = e.e^\alpha_a \delta e^a_\alpha$$

$$= c_4 e.e^\alpha_a \bar{\epsilon}(x)\rho^a \chi_\alpha(x)$$

$$= c_4 \bar{\epsilon}(x)\rho^a \chi_a = c_4 \bar{\epsilon}\rho^a \chi_a$$

where the dyad $e^a_\alpha(x)$ is used for lowering and raising the spinor indices from between the locally inertial frame and the coordinate frame.

Note: If A is a square matrix then

$$\delta det(A) = A^{ij}dA_{ij} = det(A)(A^{-1})_{ij}dA_{ij} = det(A).Tr(A^{-T}dA)$$

where $((A^{ij}))$ is the cofactor matrix of $A = ((A_{ij}))$.

Remark:If we drop the terms involving the field χ_α, ie, retain only the first term in S, then we get only global supersymmetry, not local. Note that if ∇_α denotes the covariant derivative, then,

$$\bar{\psi}^\mu \rho^\alpha \nabla_\alpha \psi_\mu = \bar{\psi}^\mu \rho^\alpha \psi_{\mu,\alpha}$$

since

$$\bar{\psi}^\mu \rho^\alpha \psi_\mu = \psi^{\mu T} \rho^0 \rho^\alpha \psi_\mu$$

and

$$\rho^0 \rho^0 = I_2, \rho^0 \rho^1 = i\sigma_2.\sigma_1 = \sigma_3$$

are symmetric matrices and therefore since the $\psi^{\mu'}s$ anticommute,

$$\psi^{\mu T} \rho^0 \rho^\alpha \psi_\mu = 0, \alpha = 0, 1$$

We shall, in what follows prove explicitly local supersymmetry of the above action under an appropriate choice of the constants c_1, c_2, c_3. We have

$$\delta(X^\mu_{,\alpha} X^{\mu,\alpha}) = 2(\delta X^\mu)_{,\alpha} X^{\mu,\alpha}$$

$$= 2(\bar{\psi}^\mu_{,\alpha}\epsilon + \bar{\psi}^\mu \epsilon_{,\alpha})X^{\mu,\alpha}$$

$$e\delta(\bar{\psi}^\mu \rho^\alpha \psi_{\mu,\alpha}) = e\delta\psi^\mu \rho^\alpha \psi_{\mu,\alpha} + e\bar{\psi}^\mu \rho^\alpha \delta\psi_{\mu,\alpha}$$

$$= -e(X^\mu_{,\alpha} - \bar{\psi}^\mu \chi_{,\alpha})\epsilon^T \rho^{\alpha T} \rho^0 \rho^\beta \psi_{\mu,\beta}$$

$$-e\psi^T_\mu \rho^0 \rho^\beta \rho^\alpha (\epsilon(X^\mu_{,\alpha} - \bar{\psi}^\mu \chi_\alpha)_{,\beta}$$

$$= -e(X^\mu_{,\alpha} - \bar{\psi}^\mu \chi_{,\alpha})\epsilon^T \rho^{\alpha T} \rho^0 \rho^\beta \psi_{\mu,\beta}$$

$$+(e\psi^T_\mu)_{,\beta} \rho^0 \rho^\beta \rho^\alpha (\epsilon(X^\mu_{,\alpha} - \bar{\psi}^\mu \chi_\alpha)$$

One of the terms in the expansion of the above is

$$e_{,\beta} \psi^T_\mu \rho^0 \rho^\beta \rho^\alpha (\epsilon(X^\mu_{,\alpha} - \bar{\psi}^\mu \chi_\alpha) - - - (1)$$

This consists of two terms:

$$e_{,\beta} \psi^T_\mu \rho^0 \rho^\beta \rho^\alpha (\epsilon\bar{\psi}^\mu \chi_\alpha) - - - (1a)$$

and

$$e_{,\beta} \psi^T_\mu \rho^0 \rho^\beta \rho^\alpha \epsilon.X^\mu_{,\alpha} - - - (1b)$$

To cancel out these terms, we must use the transformation $\delta\chi_\alpha = \nabla_\alpha \epsilon$ in the component Lagrangian involving c_3 and in another term involving c_2. Specifically, the term (1a) is cancelled by one of the terms in

$$c_3 \delta_\chi (e(x) \bar{\psi}_\mu \psi^\mu \bar{\chi}_\alpha \rho^\beta \rho^\alpha \chi_\beta)$$

$$= c_3 e(x)\bar{\psi}_\mu \psi^\mu (\nabla_\alpha \bar{\epsilon})\rho^\beta \rho^\alpha \chi_\beta$$
$$+c_3 e(x)\bar{\psi}_\mu \psi^\mu \bar{\chi}_\alpha \rho^\beta \rho^\alpha \nabla_\beta \epsilon$$
$$= c_3 \bar{\psi}_\mu \psi^\mu (e(x)\bar{\epsilon})_{,\alpha} \rho^\beta \rho^\alpha \chi_\beta$$
$$+c_3 \bar{\psi}_\mu \psi^\mu \bar{\chi}_\alpha \rho^\beta \rho^\alpha (e(x)\epsilon(x))_{,\beta} --- (2a)$$

involving the partial derivative of e but not of ϵ (we've made use of the well known identity (in general relativity) involving covariant derivatives

$$e(x).\nabla_\alpha \epsilon(x) = (e(x)\epsilon(x))_{,\alpha}$$

$$e(x).\nabla_\alpha \bar{\epsilon}(x) = (e(x)\bar{\epsilon}(x))_{,\alpha}$$

) The term (1b) is cancelled by one of the terms in

$$c_2 \delta_\chi (e(x)\bar{\chi}_\alpha \rho^\beta \rho^\alpha \psi^\mu X_{\mu,\beta}$$

$$= c_2 e(x)(\nabla_\alpha \bar{\epsilon}(x))\rho^\beta \rho^\alpha \psi^\mu X_{\mu,\beta}$$
$$= c_2 (e(x)\bar{\epsilon}(x))_{,\alpha} \rho^\beta \rho^\alpha \psi^\mu X_{\mu,\beta} --- (2b)$$

involving partial derivative of e but not of ϵ. The sum of the two terms in (2a) involving partial derivative of ϵ but not of e has two ψ factors and one χ factor without derivatives evaluate to $c_3 \bar{\psi}_\mu \psi^\mu$ times

$$e\bar{\epsilon}_{,\alpha} \rho^\beta \rho^\alpha \chi_\beta$$

$$+e\bar{\chi}_\alpha \rho^\beta \rho^\alpha \epsilon_{,\beta}$$

$$= e\epsilon^T_{,\alpha} \rho^0 \rho^\beta \rho^\alpha \chi_\beta$$

$$+e\chi^T_\beta \rho^0 \rho^\alpha \rho^\beta \epsilon_{,\alpha}$$

$$= e\chi^T_\beta \rho^{\alpha T} \rho^{\beta T} \rho^0 \epsilon_{,\alpha}$$

$$+e\chi^T_\beta \rho^0 \rho^\alpha \rho^\beta \epsilon_{,\alpha}$$

$$= e\chi^T_\beta \rho^0 \rho^\alpha \rho^\beta \epsilon_{,\alpha}$$

In other words, this term is

$$2e\bar{\psi}_\mu \psi^\mu \bar{\chi}_\beta \rho^\alpha \rho^\beta \epsilon^{,\alpha} --- (3)$$

We now look at the variation of the Lagrangian component involving c_2 w.r.t. X^μ. This variation is

$$c_2 \delta_X (e(x)\bar{\chi}_\alpha \rho^\beta \rho^\alpha \psi^\mu X_{\mu,\beta})$$
$$= c_2 e(x)\bar{\chi}_\alpha \rho^\beta \rho^\alpha \psi^\mu \delta X_{\mu,\beta}$$
$$= c_2 e(x)\bar{\chi}_\alpha \rho^\beta \rho^\alpha \psi^\mu (\bar{\epsilon}\psi_\mu)_{,\beta}$$

One of the terms in this is

$$c_2 e(x)\bar{\chi}_\alpha \rho^\beta \rho^\alpha \psi^\mu \bar{\epsilon}_{,\beta} \psi_\mu$$

$$= c_2 e(x)\bar{\chi}_\alpha \rho^\beta \rho^\alpha \psi^\mu \bar{\psi}_\mu \epsilon_{,\beta} - - - (4)$$

Now, $\psi^\mu \psi_\mu^T$ is a skewsymmetric 2×2 matrix and so it is proportional to $\rho^0 = \sigma_2$. So we can write

$$\psi^\mu \bar{\psi}_\mu = \psi^\mu \psi_\mu^T \rho^0$$

$$= (\psi^{\mu T} \rho^0 \psi_\mu \rho^0)\rho^0 = \bar{\psi}^\mu \psi_\mu . I_2$$

$$= \bar{\psi}_\mu \psi^\mu . I_2$$

and therefore (4) can be made to cancel with (3).

We now look at the variation in the first term of the component involving c_1

[17] Relationship between quantum string theory and the transmission of messages in our nervous system; Each nerve is a long pipe at the ends of which are attached synapse dendrites like the branches of a tree. In quantum string theory, one computes the amplitude

$$< \phi_2|\Delta.V(k_1, z_1)\Delta.V(k_2, z_2)...V(k_N, z_N)\Delta|\phi_1 >$$

where Δ is the string propagator and $V(k, z)$ is a vertex function defined by

$$V(k, z) =: exp(ik.X(z)) := exp(k.\sum_{n<0} \alpha(n)z^n/n).exp(k.\sum_{n>0}(\alpha(n)/n)z^n/n)$$

where $\alpha^\mu(n)$ are the Bosonic creation and annihilation operators satisfying the CCR

$$[\alpha^\mu(n), \alpha^\nu(m)] = n\eta^{\mu\nu}.\delta(n+m)$$

This quantum mechanical amplitude for example can be used to construct, after an appropriate weighted integration over $z_1, ..., z_N$, a typical term in the scattering matrix amplitude between the initial state $|\phi_1 >$ and the final state $|\phi_2 >$ when external lines carrying momenta $k_1, ..., k_N$ are attached to the string at different points. It can also be used to describe a typical term in the scattering of incoming particles by a string with initial momenta being some subset of $k_1, ..., k_N$ and final momenta the complement of this set. The string propagator is essentially $(L_0 - 1)^{-1}$ where

$$L_0 = p^2/2 + \sum_{n>0} \alpha(-n).\alpha(n)$$

is the string Hamiltonian. This is just as in conventional quantum field theory where the Klein-Gordon propagator is given by $(p^2 - m^2 + i\epsilon)^{-1}$ with $p^2 = p^\mu p_\mu$ and m being the mass of the Klein-Gordon particle. In this case of Bosonic strings, the mass shell is defined by $p^2 = 2$ or equivalently by $p^2/2 = 1$ while the particles associated with the $\alpha(n)'s$ just like photons have zero mass. The propagator for Fermionic fields is different. In conventional quantum field theory, the electron propagator is $(\gamma.p - m + i\epsilon)^{-1}$ which is the same as $(\gamma.p+m)/(p^2-m^2+i\epsilon)$. In the case of a Fermionic string, the propagator becomes $F_0^{-1} = F_0/G_0$ where

$$F_0 = \sum_n S_{-n}.\alpha(n), G_0 = (1/2)\sum_n(\alpha(-n).\alpha(n) + nS_{-n}.S_n)$$

We note that

$$F_0^2 = \sum_{n,m} S_{-n}.\alpha(n)S_{-m}.\alpha(m)$$

$$= \sum_{n,m} S^{\mu}_{-n}S^{\nu}(-m)\alpha_{\mu}(n)\alpha_{\nu}(m)$$

$$= \sum_{n,m} S^{\mu}_{-n}S^{\nu}_{-m}([\alpha_{\mu}(n), \alpha_{\nu}(m)] + \alpha_{\nu}(m)\alpha_{\mu}(n))$$

$$= \sum_{n,m} S^{\mu}_{-n}S^{\nu}_{-m}(n\eta_{\mu\nu}\delta(n+m) + \alpha_{\nu}(m)\alpha_{\mu}(m))$$

$$= \eta_{\mu\nu}\sum_n nS^{\mu}_{-n}S^{\nu}_n$$

$$+\sum_{n,m} S^{\mu}_{-n}S^{\nu}(-m)\alpha_{\nu}(m)\alpha_{\mu}(m)$$

$$= \eta_{\mu\nu}\sum_n nS^{\mu}_{-n}S^{\nu}_n$$

$$+\sum_{n,m}(\{S^{\mu}_{-n}, S^{\nu}(-m)\} - S^{\nu}_{-m}S^{\mu}_{-n})\alpha_{\nu}(m)\alpha_{\mu}(m)$$

$$= \sum_n nS_{-n}.S_n$$

$$+\sum_{n,m}(\eta^{\mu\nu}\delta(n+m) - S^{\nu}(-m)S^{\mu}(-n))\alpha_{\nu}(m)\alpha_{\mu}(m)$$

$$= \sum_n nS_{-n}.S_n + \sum_n \alpha(-n).\alpha(n) - F_0^2$$

Thus,

$$F_0^2 = (1/2)(\sum_n S_{-n}.S_n + \alpha(-n).\alpha(n)) = G_0$$

Note that G_0 represents the total energy in the Bosonic and Fermionic components of the superstring.

[18] **Supersymmetric fluid dynamics**

The velocity field $v^{\mu}(x,\theta)$ is a super vector field. x are the space-time coordinates and θ are the Majorana Fermionic coordinates. The rate of change of the velocity field with time is given by

$$dv^{\mu}/dt = v^{\mu}_{,nu}(x,\theta)v^{\nu}(x,\theta) + v^{\mu}_{,a}\theta^{a'}$$

where

$$v^{\mu}_{,nu} = \partial v^{\mu}/\partial x^{\nu}, v^{\mu}_{,a} = \partial v^{\mu}/\partial \theta^{a}$$

In this expression, for a point particle in the fluid, we have

$$v^{\mu} = dx^{\mu}/dt = x^{\mu'}$$

Likewise, we define

$$u^{a} = d\theta^{a}/dt = \theta^{a'}$$

so that

$$du^{a}/dt = u^{a}_{,nu}u^{\nu} + u^{a}_{,b}u^{b}$$

Also,

$$dv^{\mu}/dt = x^{\mu''}, du^{a}/dt = \theta^{a''}$$

We substitute for $x^{\mu''}$ and $\theta^{a''}$ their expressions derived for a point particle above and add external super-force fields to the equations of motion.

[19] Supersymmetric Linear Algebra:
The super-algebra $\mathcal{A}$ is $\mathbb{Z}_2$ graded, ie, we can write it as

$$\mathcal{A} = \mathcal{A}_0 \oplus \mathcal{A}_1$$

where

$$\mathcal{A}_k.\mathcal{A}_m \subset \mathcal{A}_{k+mmod2}$$

or in full expanded form

$$\mathcal{A}_0.\mathcal{A}_0 \subset \mathcal{A}_0,$$

$$\mathcal{A}_1.\mathcal{A}_1 \subset \mathcal{A}_0,$$

$$\mathcal{A}_0.\mathcal{A}_1 \subset \mathcal{A}_1,$$

$$\mathcal{A}_1.\mathcal{A}_0 \subset \mathcal{A}_1$$

We define

$$p(x) = 0, x \in \mathcal{A}_0, p(x) = 1, x \in \mathcal{A}_1$$

Then, we introduce a super Lie bracket in $\mathcal{A}$ by

$$\{X, Y\} = XY - (-1)^{p(X)p(Y)}YX, forX, Y \in \mathcal{A}_0 \bigcup \mathcal{A}_1$$

and then extend this definition bilinearly to cover all $X, Y \in \mathcal{A}$. Then $\mathcal{A}$ with this bracket becomes a super-Lie algebra.

Exercise: What is the Jacobi identity for a super-Lie algebra ?

[20] Introducing Fermionic parameters and variables into ordinary differential equations, partial differential equations, stochastic differential equations and quantum stochastic differential equations.

[a] For the motion of a point particle in general relativity, we wrote down an action functional by incorporating Fermionic variables along with Bosonic variables. Make Newtonian approximations like $e(x) = 1+U(x)$ where $|U(x)| << 1$

and low velocities, ie, $dx^0/dt = 1 >> |dx^r/dt|, r = 1,2,3$ and derive the resulting supersymmetric version of the Newtonian differential equations of mechanics

$$\frac{d^2x^r}{dt^2} = -\frac{\partial U(x)}{\partial x^r}, r = 1,2,3$$

hint: The equations of motion are

$$e(x)(x^{\mu''} + \theta^T \epsilon\Gamma^\mu\theta'' + \theta^{'T}\epsilon\Gamma^\mu\theta'))+$$

$$e_{,\nu}(x)x^{\nu'}(x^{\mu'} + \theta^T\epsilon\Gamma^\mu\theta')$$

$$= (1/2)e^{,\mu}(x)(x^{\nu'} + \theta^T\epsilon\Gamma^\nu\theta')^2,$$

Making cancellations in the θ equation gives

$$e(x)\epsilon\Gamma^\mu\theta.p'_\mu = e_{,\nu}(x)x^{\nu'}\epsilon\Gamma^\mu\theta.p_\mu$$

Taking $\mu = r = 1,2,3$, the above equations in the non-relativistic approximation become when $e(x)$ is time independent (ie, the Newtonian gravitational potential is time independent),

$$x^{r''} + \theta^T\epsilon\Gamma^r\theta'' + \theta^{'T}\epsilon\Gamma^r\theta'$$

$$= (-1/2)e_{,r}p^2,$$

$$e(x)\epsilon(\Gamma^0\theta.p'_0 + \Gamma^r\theta.p'_r) = e_{,s}x^{s'}\epsilon(\Gamma^0\theta.p_0 + \Gamma^r\theta.p_r)$$

Assuming that the particle has unit mass $p^2 = 1$ from special relativity assuming that the Fermionic contribution to p^μ is negligible and hence we obtain a further approximation to the equations of motion:

$$v^{r'} + \theta^T\epsilon\Gamma^r\theta'' + \theta^{'T}\epsilon\Gamma^r\theta' = -U_{,r}(x),$$

$$\Gamma^\mu[p'_\mu - e_{,s}v^s p_\mu]\theta = 0$$

where

$$v^r = x^{r'}$$

It is not clear how to obtain closed form solutions to these approximate equations of motion. If we can find a constant Fermionic parameter vector θ for which

$$\Gamma^\mu\theta = 0$$

Then we get back to the ordinary Newtonian equations of motion

$$x^{r''} = v^{r'} = -U_{,r}(x)$$

However, the second, ie, the θ equation will also be satisfied if

$$\Gamma^\mu[p'_\mu - e_{,s}v^s p_\mu] = 0$$

and the first equation, ie, the x-equation is also expressible as

$$\frac{d}{dt}(v^r + \theta^T \epsilon \Gamma^r \theta') = -U_{,r}(x)$$

[21] **Fermionic coherent states**

Let γ be a Fermionic parameter, ie, $\gamma^2 = 0$ and γ anticommutes with γ^* and also with a, a^* where

$$a^2 = 0, \{a, a^*\} = 1$$

Let $|0>$ be the vacuum so that

$$a|0>= 0, a^*|0>= 1, a^{*2} = 0, a^*|1>= 0$$

Consider a state

$$|\gamma>= (p\gamma + q)|0> +(c\gamma + d)|1>, p, q, c, d \in \mathbb{C}$$

We wish that

$$a|\gamma>= \gamma|\gamma>$$

Then we should have

$$q\gamma|0> +d\gamma|1>= (d - c\gamma)|0>$$

Thus,

$$d = 0, q = -c$$

and hence

$$|\gamma>= (p\gamma - c)|0> +c\gamma|1>$$

This state is called a Fermionic coherent state.

[22] Campbell-Baker-Hausdorff formula and its application to the construction of non-Abelian gauge supersymmetric Lagrangians.

Let X, Y be two non-commuting matrices of the same size. Then we write

$$Z(t) = log(exp(tX).exp(tY))$$

and derive using the well known formula for the differential of the exponential map in Lie algebra theory,

$$exp(tX)(X + Y)exp(tY) = \frac{d}{dt}exp(Z(t)) =$$

$$exp(Z(t))(I - exp(-ad(Z(t)))/ad(Z(t))(Z'(t))$$

writing

$$g(z) = (1 - exp(-z))/z$$

we have

$$g(adZ(t))(Z'(t)) = exp(-Z(t))exp(tX)(X+Y)exp(tY)$$

$$= exp(-tY)(X+Y)ex(tY) = exp(-t.ad(Y))(X) + Y$$

or equivalently,

$$Z'(t) = g(ad(Z(t))^{-1}(exp(-t.ad(Y))(X){+}Y)$$
$$= g(ad(Z(t))^{-1}(X{+}Y{+}(exp(-t.ad(Y)){-}I)(X))$$

By Taylor expanding $g(z)^{-1}$ in powers of z, this equation can be recognized as a differential equation for $Z(t)$ with a power series solution of the form

$$Z(t) = t(X+Y) + c_2(X,Y)t^2 + c_3(X,Y)t^3 + ..$$

where $c_k(X,Y)$ is built out of multiple commutators of X and Y.

This formula finds its application to supersymmetry in the following context. Consider the restricted gauge transformation of the gauge field $\Gamma(x,\theta) = exp(t.V) = exp(t^A V_A(x,\theta))$ given by

$$\Gamma(x,\theta) \rightarrow exp(it^A\Lambda_A(x_+))\Gamma(x,\theta)exp(-it^A\Lambda_A(x_+)^*)$$

We require to evaluate the component fields of this superfield in order to derive explicit forms of the gauge transformations of the gauge field $V_\mu^A(x)$, the gaugino field $\lambda^A(x)$ and the auxiliary field $D^A(x)$ appearing as components of the gauge superfield $V^A(x,\theta)$. In the Wess-Zumino gauge, only these three components appear in V^A. Any gauge superfield can be brought into the Wess-Zumino gauge via an extended gauge transformation, ie by replacing the left Chiral field $\Lambda_A(x_+)$ by an arbitary left Chiral field $\Omega_A(x_+,\theta_L)$. After the gauge superfield has been expressed in the Wess-Zumino gauge, it remains in the same gauge under ordinary gauge transformations but not under extended gauge transformations. Therefore it becomes important to calculate the change in the components of the gauge superfield V_A under ordinary gauge transformations when $\Lambda_A(x_+)$ is infinitesimal. This involves expressing

$$exp(it^A\Lambda_A(x_+))exp(t^A V_A(x,\theta))exp(-it^A\Lambda_A(x_+)^*)$$

as the exponential of another superfield with the exponent being computed only upto linear orders in $\Lambda_A(x)$ and to evaluate this, we require the Baker-Campbell-Hausdorff formula.

Chapter 7

The Atiyah-Singer Index Theorem and Its Application to Anomalies in Quantum Field Theory

Historic remark: V.K.Patodi was an Indian mathematician who refined the Atiyah-Singer index theorem to include manifolds with a boundary. The index of the Dirac operator on such a manifold given by the formula of Atiyah and Singer acquires an additional boundary term which formed the main core of Patodi's work.

An elementary supersymmetric harmonic oscillator based proof of the Atiyah-Singer Index theorem for Dirac operators in the superposition of a gravitational field an a Yang-Mills non-Abelian gauge fieldby Harish Parthasarathy
ECE division, NSUT

I shall not be too rigorous but shall give a proof of this celebrated theorem from the standpoint of the theoretical physicist who is interested more in using the Index theorem for characterizing anomalies, namely violation of current conservation caused by Chiralities and non-Abelian transformations which cause the path measure in field space to be no longer invariant.

Let $\gamma^a, a = 1, 2, ..., N$ denote the constant flat Minkowski space-time Gamma matrices satisfying the Dirac anticommutation relations:

$$\{\gamma^a, \gamma^b\} = 2\eta^{ab}$$

where

$$((\eta^{ab})) = diag[1, -1, ..., = -1]$$

is the Minkowski metric of flat space-time. Let $g^{\mu\nu}(x)$ be the metric of curved space-time and $e^\mu_a(x)$ a tetrad basis, ie, a locally inertial frame. We raise and lower the Greek indices like $\mu, \nu, \rho, \sigma, \alpha, \beta$ using the metrics $g^{\mu\nu}$ and $g_{\mu\nu}$ or curved space-time and raise and lower the Roman indices like a, b, c, d using the Minkowski metric $((\eta^{ab})), (\eta_{ab} = \eta^{ab}))$ of flat space time. We thus have the following consistent set of matrix identities:

$$g^{\mu\nu}(x) = \eta^{ab} e_a^\mu(x) e_b^\nu(x)$$

$$e_{a\mu}(x) = g_{\mu\nu}(x) e_a^\nu(x) = \eta_{ab} e_\mu^b(x),$$

$$e_a^\mu(x) = g^{\mu\nu}(x) e_{a\nu}(x) = \eta_{ab} e^{b\mu}(x)$$

$$g_{\mu\nu}(x) g^{\nu\rho}(x) = \delta_\mu^\rho,$$

$$\eta_{ab}\eta^{bc}\delta_a^c,$$

$$g_{\mu\nu}(x) e_a^\mu(x) e_b^\nu(x) = e_{a\mu}(x) e_b^\mu(x) = \eta_{ab}, e_{a\mu}(x) e_\nu^a(x) = g_{\mu\nu}(x),$$

$$g_{\mu\nu}(x) e_a^\mu(x) e_b^\nu(x) = \eta_{ab},$$

We define the curved space-time non-constant Dirac matrices $\gamma^\mu(x)$ by

$$\gamma^\mu(x) = \gamma^a e_a^\mu(x)$$

It is easy to see from the anticommutation relations for the flat space-time Gamma matrices and the above properties of the tetrad frame that

$$\{\gamma^\mu(x), \gamma^\nu(x)\} = g^{\mu\nu}(x)$$

Let $\Gamma_\mu(x)$ denote the spinor connection of the gravitational field, ie, if $\Lambda(x)$ is a local Lorentz transformation, (ie, $\eta_{ab}\Lambda_c^a(x)\Lambda_d^b(x) = \eta_{cd}$), and if $\Lambda \to D(\Lambda)$ denotes the spinor representation of the Lorentz group, then

$$D(\Lambda(x))(\partial_\mu + \Gamma_\mu(x)) D(\Lambda(x))^{-1} =$$

$$= \partial_\mu + \Gamma_{\mu'}(x)$$

or equivalently,

$$\Gamma'_\mu(x) = (D(\Lambda(x))(\partial_\mu D(\Lambda(x))^{-1}) + D(\Lambda(x))\Gamma_\mu(x).D(\Lambda(x))^{-1}$$

defines the transformation of the spinor connection of the gravitational field under local Lorentz transformations. Such a connection can be constructed as

$$\Gamma_\mu(x) = (1/2) e_{a\nu:\mu}(x) e_b^\nu(x) J^{ab}$$

where

$$J^{ab} = (1/4)[\gamma^a, \gamma^b]$$

is the canonical basis for the Lie algebra of the spinor representation of the Lorentz group. Specifically, the canonical basis ω_{ab} for Lorentz transformations of space-time has matrix elements given by

$$[omega_{ab}]_{\mu\nu} = \eta_{a\mu}\eta_{b\nu} - \eta_{a\nu}\eta_{n\mu}$$

or equivalently by raising and lowering indices by

$$[\omega_{ab}]^{\mu}_{\nu} = \delta^{\mu}_{a}\eta_{b\nu} - \eta_{a\nu}\delta^{\mu}_{b}$$

$$[\omega_{ab}]^{\mu\nu} = \eta^{a\mu}\eta^{b\nu} - \eta^{a\nu}\eta^{b\mu}$$

Note that the multiplication of two such canonical Lorentz transformations is given by

$$[\omega_{ab}\omega_{cd}]^{\mu}_{\nu} = [\omega_{ab}]^{\mu}_{\alpha}[\omega_{cd}]^{\alpha}_{\nu}$$

The action of such a canonical Lorentz transformation on a contravariant vector A^{μ} is given by

$$A^{'\mu} = [\omega_{ab}]^{\mu}_{\nu}A^{\nu}$$

and likewise, we have since

$$[\omega_{ab}]_{\mu\nu} = \eta_{\mu\alpha}[\omega_{ab}]^{\alpha}_{\nu}$$

etc., that

$$A'_{\mu} = \eta_{\mu\nu}A^{'\nu} = [\omega_{ab}]_{\mu\nu}A^{\nu}$$

It is easily verified that

$$A'_{\mu}B^{'\mu} = A_{\mu}B^{\mu}$$

The structure constants of the Lorentz group corresponding to the six basis elements $\omega_{ab}, 0 \leq a < b \leq 3$ are easily computed using

$$[\omega_{ab}, \omega_{cd}] = \eta_{ad}\omega_{bc} + \eta_{bc}\omega_{ad} - \eta_{ac}\omega_{bd} - \eta_{bd}\omega_{ac}$$

We have that

$$J^{ab} = dD(\omega^{ab})$$

where

$$\omega^{ab} = \eta^{ac}\eta^{bd}\omega_{cd}$$

ie, the basis J^{ab} for the spinor representation of the Lorentz group is obtained by acting on the canonical Lorentz generators ω^{ab} of space-time transformations by the differential of the spinor representation $D(.)$ of the Lorentz group. In fact, this is actually the way in which the spinor representation D of the Lorentz group is defined, once we verify that

$$[dD(\omega^{ab}), dD(\omega^{cd})] = [J^{ab}, J^{cd}] =$$

$$dD([\omega^{ab}, \omega^{cd}]) = dD(\eta^{ad}J^{bc} + \eta^{bc}J^{ad} - \eta^{ac}\omega^{bd} - \eta^{bd}\omega^{ac})$$

$$= \eta^{ad}dD(\omega^{bc}) + \eta^{bc}dD(\omega^{ad}) - \eta^{ac}dD(\omega^{bd}) - \eta^{bd}dD(\omega^{ac})$$

$$= \eta^{ad}J^{bc} + \eta^{bc}J^{ad} - \eta^{ac}J^{bd} - \eta^{bd}J^{ac}$$

This commutation relation for $J^{ab} = (1/4)[\gamma^{a}, \gamma^{b}]$ is easily verified by writing

$$[\gamma^{a}, \gamma^{b}] = 2(\gamma^{a}\gamma^{b} - \eta^{ab})$$

and using anticommutation properties of the Dirac matrices. In other words, we get a valid spinor representation of the Lorentz Lie algebra once we use the fact that any representation of a Lie algebra must preserve the structure constants.

The gravitational spin connection Γ_μ constructed above can easily be shown to satisfy the above transformation property under local Lorentz transformations. The method to prove this is to adopt the infinitesimal approach. Let $\Lambda(x) = I + \omega(x)$ be an infinitesimal local Lorentz transformation. Then under this transformation, $\Gamma_\mu(x) = J^{ab} e_{a\nu:\mu} e_b^\nu$ transforms to

$$\Gamma'_\mu(x) = J^{ab}(\Lambda_a^c(x) e_{c\nu}(x))_{:\mu} \Lambda_b^d(x) e_d^\nu(x)$$

$$= \Gamma_\mu(x) + J^{ab}[\omega_{a,\mu}^c(x) e_{c\nu}(x) e_d^\nu(x) + e_{a\nu:\mu}(x) \omega_b^d(x) d_d^\nu(x)]$$

with neglect of $O(|\omega|^2)$ terms. Note that in this expression, $\omega_a^c = \eta^{cd}\omega_{ad} = -\eta^{cd}\omega_{da}$. On the other hand, we wish that this coincide with the expression

$$D(I + \omega(x))\Gamma_\mu(x) D(I + \omega(x))^{-1} + D(I + \omega(x))\partial_\mu D(I + \omega(x))^{-1}$$

$$= \Gamma_\mu(x) + [dD(\omega(x)), \Gamma_\mu(x)] - \partial_\mu dD(\omega(x))$$
$$= \Gamma_\mu(x) + \omega_{ab}(x)[J^{ab}, \Gamma_\mu(x)] - \omega_{ab,\mu}(x) J^{ab}$$

where

$$[J^{ab}, \Gamma_\mu(x)] = [J^{ab}, J^{cd}] e_{c\nu:\mu}(x) e_{d:\mu}^\nu(x)$$

and this can be shown to be the case by using the expression for the commutator $[J^{ab}, J^{cd}]$. Note that $\Lambda_b^a(x)$ are diffeomorphism scalars since they act on the inertial frame indices a, b, c etc in e_a^μ, e_μ^a etc.

Getzler scaling: We can define the Dirac Gamma matrices using the Fermionic creation and annihilation operators on the antisymmetric tensor algebra of a vector space. Let V be a vector space of dimension n and let

$$\Lambda V = \bigoplus_{k=0}^{n} \Lambda^k V$$

Then ΛV is a 2^n-dimensional vector space and we choose a basis $\{e^a, a = 1, 2, ..., n\}$ for this vector space. We also denote by $\{e_a, a = 1, 2, ..., n\}$ the dual of this basis:

$$(e^a, e_b) = e_b(e^a) = \delta_b^a$$

Then define the action of e^a on ΛV by

$$e^a.\omega = e^a \wedge \omega, \omega \in \Lambda V$$

Also define the action of e_a on ΛV as

$$e_a(v^1 \wedge ... \wedge v^r) = \sum_{k=1`}^{r} (-1)^{k-1}(e_a, v^k)(v^1 \wedge ... \wedge \hat{v}^k \wedge ... \wedge v^k)$$

It is then easy to verify that for this action,

$$e^a e_b + e_b e^a = \delta^a_b$$

and that in fact e_a is the adjoint operator of e^a in ΛV, ie,

$$(e^a \omega, \theta) = (\omega, e_a \theta)$$

Now define the Dirac operators on ΛV by

$$\gamma^a = e^a + \eta^{ab} e_b$$

Then using the obvious fact that

$$e^a e^b + e^b e^a = 0, e_a e_b + e_b e_a = 0, e^a e_b + e_b e^a = \delta^a_b$$

we easily deduce that

$$\gamma^a.\gamma^b + \gamma^b.\gamma^a = (e^a + \eta^{ac} e_c)(e^b + \eta^{bd} e_d) + (e^b + \eta^{bd} e_d)(e^a + \eta^{ac} e_c) = 2\eta^{ab}$$

Note that we could define additional Dirac matrices

$$\gamma^{n+a} = e^a - \eta^{ab} e_b$$

and we would then get a set of $2n$ Dirac matrices $\gamma^a, a = 1, 2, ..., n$ satisfying the standard anticommutation relations. Now for $\epsilon > 0$ with ϵ to be interpreted as a small number converging to zero, we define the operator S_ϵ acting on $\Lambda V \oplus \Lambda V^*$ as

$$S_\epsilon \omega = \epsilon^{-\partial \omega} \omega$$

where $\partial \omega$ is the degree of ω if $\omega \in \Lambda^k V$, then we take $\partial \omega = k$, if $\omega \in \Lambda^k V^*$, we take $\partial \omega = -k$ and extend S_ϵ by linearity. We then easily verify that since the action of e^a raises the degree of a skewsymmetric tensor by one and that by e_a lowers it by one, we have

$$S_\epsilon e^a S_\epsilon^{-1} \omega = S_\epsilon e^a \epsilon^{\partial \omega} \omega$$

$$= \epsilon^{\partial \omega} S_\epsilon (e^a.\omega) = \epsilon^{\partial \omega}.\epsilon^{-\partial \omega - 1} e^a.\omega = \epsilon^{-1} e^a.\omega,$$

$$S_\epsilon e_a S_\epsilon^{-1} \omega = S_\epsilon e_a \epsilon^{\partial \omega} \omega$$

$$= \epsilon^{\partial \omega} S_\epsilon e_a.\omega = \epsilon^{\partial \omega}.\epsilon^{-\partial \omega + 1} e_a.\omega = \epsilon.e_a.\omega$$

so that

$$S_\epsilon \gamma^a.S_\epsilon^{-1} = \epsilon^{-1} e^a + \epsilon e_a$$

or equivalently,

$$\epsilon.S_\epsilon.\gamma^a.S_\epsilon^{-1} = e^a + \epsilon^2 e_a$$

The general structure of the Dirac operator: The Dirac operator acts on sections of a vector bundle given by taking a Riemannian manifold $\mathcal{M}$ and attaching to each point $x \in \mathcal{M}$ a vector space V_x which is isomorphic to a given fixed vector space V. The vector space V is the tensor product of two vector spaces V_1 and V_2. The Dirac operators γ^a and hence the spinor connection of the gravitational field act in V_1 and the Yang-Mills fields act in V_2. Specifically, we write the Dirac operator as

$$D = \gamma^\mu(x)\nabla_\mu, \nabla_\mu = \partial_\mu + \Gamma_\mu(x) + A_\mu(x)$$

or more precisely,

$$D = \gamma^\mu(x)\partial_\mu = (\gamma^\mu(x) \otimes I_{V_2})\partial_\mu,$$

$$\Gamma_\mu(x) = \Gamma_\mu(x) \otimes I_{V_2},$$

$$A_\mu(x) = I_{V_1} \otimes A_\mu(x)$$

We write

$$A_\mu(x) = A_\mu^a(x) i\tau_a$$

where τ_a are Hermitian matrices acting in V_2. $\{i\tau_a\}$ form a Lie algebra of skew-Hermitian matrices acting in V_2. In particular, we note that the operators $\Gamma_\mu(x)$ and $A_\mu(x)$ commute at every $x \in \mathcal{M}$. The square of the Dirac operator in this full generality was first obtained by Lichnerowicz. It is given by

$$D^2 = \gamma^\mu \nabla_\mu \gamma^\nu \nabla_\nu =$$

$$= \gamma^\mu\gamma^\nu\nabla_\mu\nabla_\nu + \gamma^\mu[\nabla_\mu, \gamma^\nu]\nabla_\nu$$

$$= (1/2)\gamma^\mu\gamma^\nu[\nabla_\mu, \nabla_\nu] + (1/2)\gamma^\mu\gamma^\nu\{\nabla_\mu, \nabla_\nu\}$$

$$+\gamma^\mu[\nabla_\mu, \gamma^\nu]\nabla_\nu$$

$$= (1/4)[\gamma^\mu, \gamma^\nu][\nabla_\mu, \nabla_\nu] + (1/4)\{\gamma^\mu, \gamma^\nu\}\{\nabla_\mu, \nabla_\nu\} + \gamma^\mu[\nabla_\mu, \gamma^\nu]\nabla_\nu$$

$$= (1/4)[\gamma^\mu, \gamma^\nu](R_{\mu\nu} + K_{\mu\nu}) + g^{\mu\nu}\nabla_\mu\nabla_\nu + \gamma^\mu[\nabla_\mu, \gamma^\nu]\nabla_\nu$$

Now,

$$[\nabla_\mu, \gamma^\nu] = [\nabla_\mu, \gamma^a e_a^\nu]$$

$$= [\partial_\mu + \Gamma_\mu, \gamma^a e_a^\nu]$$

$$= \gamma^a e_{a,\mu}^\nu + [\Gamma_\mu, \gamma^a]e_a^\nu$$

Thus,

$$\gamma^\mu[\nabla_\mu, \gamma^\nu] =$$

$$\gamma^b\gamma^a e_b^\mu e_{a,\mu}^\nu + [J^{bc}, \gamma^a] e_{b:\mu}^\alpha e_{c\alpha} e_a^\nu$$

Note that Γ_μ acts in V_1 (or more precisely in V_{1x}) while $A_\mu(x)$ acts in V_2 (or more precisely in V_{2x}) and hence these two operators commute at each $x \in \mathcal{M}$. Thus,

$$[\nabla_\mu, \nabla_\nu] = [\partial_\mu + \Gamma_\mu + A_\mu, \partial_\nu + \Gamma_\nu + A_\nu] =$$

$$R_{\mu\nu}(x) + K_{\mu\nu}(x)$$

where

$$R_{\mu\nu}(x) = \Gamma_{\nu,\mu} - \Gamma_{\mu,\nu} + [\Gamma_\mu, \Gamma_\nu]$$

is a $dimV_1 \times dimV_1$ matrix acting in V_{1x} while

$$K_{\mu\nu}(x) = A_{\nu,\mu} - A_{\mu,\nu} + [A_\mu, A_\nu]$$

is a $dimV_2 \times dimV_2$ matrix acting in V_{2x}. Given a differential operator L acting on the vector bundle $\bigcup_{x\in\mathcal{M}} V_x$, we define the Getzler scaling operator S_ϵ acting on this vector bundle so that it changes x to ϵx and changes a matrix $M(x)$ in $V_{1x} \times V_{1x}$ to $S_\epsilon M(\epsilon x) S_\epsilon^{-1}$. More specifically, if we write

$$L = \sum_{a\mu_1 \ldots \mu_k, k} P_{\mu_1,\ldots,\mu_k a}(x) \otimes Q_{\mu_1 \ldots \mu_k, a}(x) \partial_{\mu_1} \ldots \partial_{\mu_k}$$

where $P_{\mu_1 \ldots \mu_k, a}(x)$ is a matrix acting on V_{1x} while $Q_{\mu_1 \ldots \mu_k, a}(x)$ is a matrix acting on V_{2x}, then

$$S_\epsilon.L.S_\epsilon^{-1}$$

is the operator

$$\sum_{a\mu_1 \ldots \mu_k, k} S_\epsilon P_{\mu_1,\ldots,\mu_k a}(\epsilon x) S_\epsilon^{-1} \otimes Q_{\mu_1 \ldots \mu_k, a}(\epsilon x) \epsilon^{-k} \partial_{\mu_1} \ldots \partial_{\mu_k}$$

where we are assuming that $P_{\mu_1 \ldots \mu_k a}(x)$ is a linear combination of products of Dirac matrices of the form $\gamma^{a_1} \ldots \gamma^{a_m}$ and

$$S_\epsilon \gamma^{a_1} \ldots \gamma^{a_m} S_\epsilon^{-1} =$$

$$(\epsilon^{-1} e^{a_1} + \epsilon e_{a_1}) \ldots (\epsilon^{-1} e^{a_m} + \epsilon e_{a_m})$$

Another important point to note that if $\gamma^1, \ldots, \gamma^n$ is a complete set of Dirac matrices, then the supertrace str of all the products $\gamma^{a_1} \ldots \gamma^{a_k}$ with distinct $a_1, \ldots, a_k$ is zero unless $k = n$ and $(a1, \ldots, an)$ is a permutation of $(1, 2, \ldots, n)$. In this case, we have

$$str(\gamma^{a1} \ldots \gamma^{an}) = c(n) sgn(a)$$

where $c(n)$ is a constant. If Equivalently, if we apply the Getzler scaling to these Dirac matrices, we then get that

$$str(\gamma^{a1} \ldots \gamma^{ak}) = \epsilon^{-n} times$$

the coefficient of $e^1 \wedge \ldots \wedge e^n$ in $S_\epsilon \gamma^{a1} \ldots \gamma^{ak} S_\epsilon^{-1}$. This is because

$$S_\epsilon \gamma^{a1} \ldots \gamma^{ak} S_\epsilon^{-1} = \epsilon^{-k} (e^{a1} + \epsilon^2 e_{a1}) \ldots (e^{ak} + \epsilon^2 e_{ak})$$

with the multiplication being taken in the exterior algebra of ΛV.

Normal coordinates: It is possible to choose locally around $x = 0$ a coordinate system such that

$$g_{\mu\nu}(x) x^\nu = \eta_{\mu\nu} x^\nu$$

This is called a normal coordinate system. In such a system, we then get on repeated differentiation,

$$g_{\mu\nu,\alpha}(x)x^\nu + g_{\mu\alpha}(x) = \eta_{\mu\alpha},$$

$$g_{\mu\rho,\alpha}(x) + g_{\mu\nu,\alpha\rho}(x)x^\nu + g_{\mu\alpha,\rho}(x) = 0$$

$$g_{\mu\rho,\alpha\nu}(0) + g_{\mu\nu,\alpha\rho} + g_{\mu\alpha,\rho\nu}(0) = 0$$

This last identity is the same as

$$\sum_{[\rho\alpha\nu]} g_{\mu\rho,alpha\nu}(0) = 0$$

where square bracket denotes cyclic sum. Note that the previous equations also imply that

$$g_{\mu\alpha}(0) = \eta_{\mu\alpha},$$

$$g_{\mu\rho,\alpha}(0) + g_{\mu\alpha,\rho}(0) = 0$$

and hence,

$$g_{\mu\rho,\alpha}(0) + g_{\rho\alpha,\mu}(0) = 0$$

from which we deduce that

$$g_{\mu\rho,\alpha}(0) = 0$$

and hence, for the Christoffel symbols,

$$\Gamma_{\mu\rho\alpha}(0) = 0$$

It follows that $g_{\mu\nu,\alpha}(x) = O(x)$ and hence $e_{a\mu,\nu}(x) = O(x)$. Thus,

$$\Gamma_\mu(x) = O(x)$$

and therefore,

$$[\Gamma_\mu(x), \Gamma_\nu(x)] = O(|x|^2)$$

Thus,

$$R_{\mu\nu}(x) = \Gamma_{\nu\,mu}(x) - \Gamma_{\mu,\nu}(x) + O(|x|^2)$$

Now,

$$\Gamma_\mu(x) = J^{ab} e_{a\alpha:\mu} e_b^\alpha =$$

$$= J^{ab} e_b^\alpha (e_{a\alpha,\mu} - \Gamma^\rho_{\alpha\mu} e_{a\rho})$$

from which it follows that

$$\Gamma_{\nu,\mu}(x) - \Gamma_{\mu,\nu}(x) =$$

$$J^{ab} e_b^\alpha(x) e_a^\rho(x)(\Gamma_{\rho\alpha\nu,\mu}(x) - \Gamma_{\rho\alpha\mu,\nu}(x)) + O(|x|^2)$$

$$= J^{ab}(\Gamma_{ab\nu,\mu}(0) - \Gamma_{ab\mu,\nu}(0)) + O(|x|^2)$$

(Note that $e_{a\mu,\nu}(x) = O(|x|)$).

Now we recall that the Riemann curvature tensor is given by

$$R_{\alpha\nu\rho\sigma}(x) = \Gamma_{\alpha\nu\rho,\sigma}(x) - \Gamma_{\alpha\nu\sigma,\rho}(x) + Q(x)$$

where $Q(x)$ denotes terms that are quadratic in the derivatives of the metric tensor. Thus,

$$R_{\alpha\nu\rho\sigma}(x) = \Gamma_{\alpha\nu\rho,\sigma}(0) - \Gamma_{\alpha\nu\sigma,\rho}(0) + O(|x|^2)$$

Combining this with the above identity gives us

$$\Gamma_{\nu,\mu}(x) - \Gamma_{\mu,\nu}(x) = \gamma^a\gamma^b R_{ab\nu\mu}(0) + O(|x|^2)$$

Remark: The relation

$$R_{\alpha\nu\rho\sigma}(0) = \Gamma_{\alpha\nu\rho,\sigma}(0) - \Gamma_{\alpha\nu\sigma,\rho}(0)$$

does not make it immediately apparent that it is antisymmetric in (α, ν). However, this is true in a normal coordinate system as can be seen by the following arguments:

$$\Gamma_{\alpha\nu\rho,\sigma}(0) - \Gamma_{\alpha\nu\sigma,\rho}(0) =$$

$$(1/2)[g_{\alpha\nu,\rho\sigma} + g_{\alpha\rho,\nu\sigma} - g_{\rho\nu,\alpha\sigma}$$

$$-g_{\alpha\nu,\sigma\rho} - g_{\alpha\sigma,\nu\rho} + g_{\sigma\nu,\alpha\rho}](0)$$

$$= (1/2)[g_{\alpha\rho,\nu\sigma} - g_{\rho\nu,\alpha\sigma} - g_{\alpha\sigma,\nu\rho} + g_{\sigma\nu,\alpha\rho}](0)$$

Now using the above cyclic identity for normal coordinates,

$$g_{\alpha\sigma,\nu\rho}(0) = -[g_{\alpha\nu,\rho\sigma}(0) + g_{\alpha\rho,\sigma\nu}(0)]$$

so that the above becomes

$$\Gamma_{\alpha\nu\rho,\sigma}(0) - \Gamma_{\alpha\nu\sigma,\rho}(0) =$$

$$(1/2)[2g_{\alpha\rho,\nu\sigma}(0) - g_{\rho\nu,\alpha\sigma}(0) + g_{\alpha\nu,\rho\sigma}(0) + g_{\sigma\nu,\alpha\rho}(0)]$$

Another use of the cyclic identity gives

$$g_{\alpha\nu,\rho\sigma}(0) + g_{\sigma\nu,\alpha\rho}(0)$$

$$= g_{\nu\alpha,\rho\sigma}(0) + g_{\nu\sigma,\alpha\rho}(0) = -g_{\nu\rho,\sigma\alpha}(0)$$

and hence, the above becomes

$$\Gamma_{\alpha\nu\rho,\sigma}(0) - \Gamma_{\alpha\nu\sigma,\rho}(0) =$$

$$= g_{\alpha\rho,\nu\sigma}(0) - g_{\nu\rho,\alpha\sigma}(0)$$

which is evidently antisymmetric in (α, ν).

Now,

$$\Gamma_{\mu}(x) = J^{ab} e_{a\nu:\mu} e^{\nu}_{b}$$

$$= (1/2)J^{ab}(e_{a\nu:\mu}e_b^\nu - e_{b\nu:\mu}e_a^\nu)$$
$$= (1/2)J^{ab}(e_b^\nu e_{a\nu,\mu} - e_a^\nu e_{b\nu,\mu})+$$
$$(1/2)J^{ab}(e_a^\nu e_b^\alpha - e_b^\nu e_a^\alpha)\Gamma\alpha\nu\mu$$
$$= \gamma^a\gamma^b e_a^\nu e_b^\alpha(\Gamma_{\alpha\nu\mu} - \Gamma_{\nu\alpha\mu})$$
$$+(1/2)J^{ab}(e_b^\nu e_{a\nu,\mu} - e_a^\nu e_{b\nu,\mu})$$

Now,

$$e_a^\nu(x) = \delta_a^\nu + O(|x|^2)$$

and therefore,

$$(e_b^\nu e_{a\nu,\mu} - e_a^\nu e_{b\nu,\mu})(x) = O(|x|^2)$$

Thus,

$$\Gamma_\mu(x) =$$
$$\gamma^a\gamma^b e_a^\nu e_b^\alpha(\Gamma_{\alpha\nu\mu} - \Gamma_{\nu\alpha\mu})(x) + O(|x|^2)$$
$$= \gamma^a\gamma^b(\Gamma_{ba\mu,\rho} - \Gamma_{ab\mu,\rho})(0)x^\rho + O(|x|^2)$$

Now,

$$\Gamma_{ba\mu,\rho}(0) - \Gamma_{ab\mu,\rho}(0) =$$
$$(1/2)(g_{ba,\mu\rho} + g_{b\mu,a\rho} - g_{\mu a,b\rho} - g_{ab,\mu\rho} - g_{a\mu,b\rho} + g_{\mu b,a\rho})(0)$$
$$= (1/2)(g_{b\mu,a\rho} - g_{\mu a,b\rho} - g_{a\mu,b\rho} + g_{\mu b,a\rho})(0)$$
$$= g_{\mu b,a\rho}(0) - g_{\mu a,b\rho}(0)$$

We've now already proved that

$$R_{\alpha\nu\rho\sigma}(0) = \Gamma_{\alpha\nu\rho,\sigma}(0) - \Gamma_{\alpha\nu\sigma,\rho}(0)$$
$$= g_{\alpha\rho,\nu\sigma}(0) - g_{\nu\rho,\alpha\sigma}(0)$$

Using this and the above identity, we find that

$$\Gamma_{ba\mu,\rho}(0) - \Gamma_{ab\mu,\rho}(0) = R_{ba\mu\rho}(0)$$

Thus, we find finally,

$$\Gamma_\mu(x) = \gamma^a\gamma^b R_{ab\rho\mu}(0)x^\rho + O(|x|^2)$$

Thus, we get finally, for the following approximation of the square of the Dirac operator in normal coordinates around a neighbourhood of the origin,

$$D^2 = \gamma^\mu\nabla_\mu\gamma^\nu\nabla_\nu =$$
$$= (1/4)[\gamma^\mu,\gamma^\nu](R_{\mu\nu} + K_{\mu\nu}) + g^{\mu\nu}\nabla_\mu\nabla_\nu+$$
$$\gamma^\mu[\nabla_\mu,\gamma^\nu]\nabla_\nu$$
$$= \eta^{\mu\nu}(\partial_\mu + \gamma^a\gamma^b R_{ab\rho\mu}(0)x^\rho + O(|x|^2))(\partial_\nu + \gamma^a\gamma^b R_{ab\rho\mu}(0)x^\rho + O(|x|^2)$$

$$+(1/4)[\gamma^\mu,\gamma^\nu]\gamma^a\gamma^b R_{ab\nu\mu}(0)+O(|x|^2)+(1/4)[\gamma^\mu,\gamma^\nu]K_{\mu\nu}$$
$$+\gamma^\mu[\nabla_\mu,\gamma^\nu)]nabla_\nu$$
$$=\eta^{\mu\nu}(\partial_\mu+\gamma^a\gamma^b R_{ab\rho\mu}(0)x^\rho)(\partial_\nu+\gamma^a\gamma^b R_{ab\mu\nu}(0)x^\rho)$$
$$+(1/2)\gamma^c\gamma^d\gamma^a\gamma^b R_{abcd}(0)+(1/2)\gamma^a\gamma^b K_{ab}(x)$$
$$+(\gamma^b\gamma^a e_b^\mu e_{a,\mu}^\nu+[J^{bc},\gamma^a]e_{b:\mu}^\alpha e_{c\alpha}e_a^\nu)\nabla_\nu O(|x|^2)$$
$$=\eta^{\mu\nu}(\partial_\mu+\gamma^a\gamma^b R_{ab\rho\mu}(0)x^\rho)(\partial_\nu+\gamma^a\gamma^b R_{ab\mu\nu}(0)x^\rho)$$
$$+R(0)+(1/2)\gamma^a\gamma^b K_{ab}(x)$$
$$+(\gamma^b\gamma^a e_b^\mu e_{a,\mu}^\nu+[J^{bc},\gamma^a]e_{b:\mu}^\alpha e_{c\alpha}e_a^\nu)\nabla_\nu$$
$$+O(|x|^2)$$

Now we observe that after Getzler scaling,

$$\nabla_\mu=(\partial_\mu+\gamma^a\gamma^b R_{ab\rho\mu}(0)x^\rho)\rightarrow$$
$$\nabla_\mu^\epsilon=\epsilon^{-1}\partial_\mu+\epsilon^{-2}(e^a+\epsilon^2\eta^{ac}e_c)(e^b+\epsilon^2\eta^{bd}e_d)R_{ab\rho\mu}(0)\epsilon x^\rho)$$

so that

$$\epsilon\nabla_\mu^\epsilon=\partial_\mu+e^a\wedge e^b R_{ab\rho\mu}(0)x^\rho+O(\epsilon)$$

Now,

$$[J^{bc},\gamma^a]=[\gamma^b\gamma^c,\gamma^a]=\gamma^b(2\eta^{ca}-\gamma^a\gamma^c)-\gamma^a\gamma^b\gamma^c$$
$$=2\eta^{ca}\gamma^b-2\eta^{ba}\gamma^c$$

so

$$(\gamma^b\gamma^a e_b^\mu e_{a,\mu}^\nu+[J^{bc},\gamma^a]e_{b:\mu}^\alpha e_{c\alpha}e_a^\nu)\nabla_\nu$$
$$=[\gamma^b\gamma^a e_b^\mu e_{a,\mu}^\nu+e_\alpha^a\gamma^b e_{b:\mu}^\alpha-e_{:\mu}^{a\alpha}e_{c\alpha}\gamma^c]\nabla_\nu$$
$$=[\gamma^b\gamma^a e_b^\mu e_{a,\mu}^\nu+\gamma^b(e_\alpha^a e_{b:\mu}^\alpha-e_{:\mu}^{a\alpha}e_{b\alpha})]\nabla_\nu$$
$$=[e^{a\mu}e_{a,\mu}^\nu+\gamma^b\gamma^a(e_b^\mu e_{a,\mu}^\nu-e_a^\mu e_{b,\mu}^\nu)$$
$$+\gamma^b(e_\alpha^a e_{b:\mu}^\alpha-e_{:\mu}^{a\alpha}e_{b\alpha})]\nabla_\nu$$

The second term, after Getzler scaling followed by multiplication with ϵ^2 behaves as $O(\epsilon)$ since γ^b after Getzler scaling behaves as $O(\epsilon^{-1})$, $e_\alpha^a e_{b:\mu}^\alpha-e_{:\mu}^{a\alpha}e_{b\alpha})$ after replacing x with ϵx behaves as $O(\epsilon)$ while ∇_ν after Getzler scaling becomes ∇_ν^ϵ which as we saw behaves as $O(\epsilon^{-1})$. Thus, in the limit as $\epsilon\rightarrow 0$, the Getzler scaled version of

$$(\gamma^b\gamma^a e_b^\mu e_{a,\mu}^\nu+[J^{bc},\gamma^a]e_{b:\mu}^\alpha e_{c\alpha}e_a^\nu)\nabla_\nu$$

followed by multiplication with ϵ^2 behaves when $\epsilon\rightarrow 0$ as

$$\epsilon^{-1}e^b\wedge e^a(e_b^\mu e_{a,\mu}^\nu-e_a^\mu e_{b,\mu}^\nu)(\epsilon x)(\partial_\nu+e^a\wedge e^b R_{ab\rho\mu}(0)x^\rho)$$

Now,

$$\epsilon^{-1}(e_b^\mu e_{a,\mu}^\nu-e_a^\mu e_{b,\mu}^\nu)(\epsilon x)=$$

$$(e^{\nu}_{a,bc}(0) - e^{\nu}_{b,ac})(0)x^c$$

Now recall that

$$e_{a\mu}e^a_\nu = g_{\mu\nu}$$

so that on differentiating twice, setting $x = 0$ and noting that in our normal coordinate system the first order partial derivatives of the tetrad vanish at $x = 0$, we get

$$e_{a\mu,bc}(0)\delta^a_\nu + \eta_{a\mu}e^a_{\nu,bc}(0) = g_{\mu\nu,bc}(0)$$

or

$$e_{\nu\mu,bc}(0) + e_{\mu\nu,bc}(0) = g_{\mu\nu,bc}(0)$$

We may assume that $e_{\mu\nu} = e_{\nu\mu}$ and then we get

$$g_{\mu\nu,bc}(0) = 2e_{\mu\nu,bc}(0)$$

Now recall that in our normal coordinate system, we have

$$g_{\mu\nu,bc}(0) - g_{\mu b,\nu c}(0) = R_{b\nu\mu c}(0)$$

and hence we get

$$e_{\nu a,bc}(0) - e_{\nu b,ac}(0) = (1/2)R_{ab\nu c}(0)$$

from which we easily deduce using the fact that the first order partial derivatives of the metric tensor and of the tetrad vanish at $x = 0$ that

$$(e^{\nu}_{a,bc}(0) - e^{\nu}_{b,ac})(0) = (1/2)R^{\nu}_{cab}(0)$$

Thus we finally get that the Getzler scaled version of D^2 followed by multiplication with ϵ^2 and then finally followed by taking the limit $\epsilon \to 0$ gives us

$$X = lim_{\epsilon\to 0}\epsilon^2(D^2)^\epsilon =$$

$$\eta^{\mu\nu}(\partial_\mu + e^a \wedge e^b R_{ab\rho\mu}(0)x^\rho)(\partial_\nu + e^a \wedge e^b R_{ab\mu\nu}(0)x^\rho)$$

$$+R(0) + (1/2)e^a \wedge e^b K_{ab}(0) - (1/2)(e^a \wedge e^b)R^{\nu}_{cab}(0)x^c(\partial_\nu + e^a \wedge e^b R_{ab\rho\mu}(0)x^\rho)$$

To cast this limiting Getzler scaled version of the square of the Dirac operator in a nice form for deriving the Atiyah-Singer index theorem, we first introduce the supersymmetric matrix, ie, a two form valued matrix

$$C = ((C_{\mu\nu})), C_{\mu\nu} = R_{\mu\nu ab}(0)e^a \wedge e^b$$

and then observe that

$$X = (\partial + Rx)^T.\eta.(\partial + Rx) + R(0) + K - (1/2)(Rx)^T(\partial + Rx)$$

Remark: We have noted that if γ^a and γ^b are two Dirac matrices, then under Getzler scaling, they transform to $\epsilon^{-1}e^a + \epsilon\eta^{ac}e_c$ and $\epsilon^{-1}e^b + \epsilon\eta^{bd}e_d$ respectively.

The product of these two Dirac matrices $\gamma^a\gamma^b$ transforms under Getzler scaling to

$$S_\epsilon\gamma^a\gamma^b S_\epsilon^{-1} = S_\epsilon\gamma^a S_\epsilon^{-1}.S_\epsilon\gamma^b S_\epsilon^{-1} =$$
$$(\epsilon^{-1}e^a + \epsilon.\eta^{ac}e_c).(\epsilon^{-1}e^b + \epsilon.\eta^{bd}e_d)$$

which is to be interpreted as multiplication in the exterior algebra $\Lambda V \otimes \Lambda V^*$. In fact, the very manner in which the Dirac matrices were constructed involved using multiplication of operators in the form of creation (wedge product) and annihilation (wedge contraction) on the vector space $\Lambda V \otimes \Lambda V^*$. The action of Fermionic creation and annihilation operators on vectors in this space via Wedge product and Wedge contraction provided us with the definition of multiplication of Dirac operators. Therefore, ordinary matrix products of Dirac matrices, after Getzler scaling should be interpreted in terms of wedge multiplication and wedge contraction. Therefore, when for example, we multiply R with itself, the result should be $C^2 = C \wedge C$ using the associativity of wedge multiplication and contraction or a mixture of both the operations. Hence, we have

$$X = (\partial + Cx)^T.\eta.(\partial + Cx) + R(0) + K - (1/2)(Cx)^T(\partial + Cx)$$
$$= \partial^T\eta\partial - (1/2)x^T C\eta Cx - (3/2)x^T C\eta\partial + R(0) + K$$

It should be noted that C, K are mutually commuting supersymmetric matrices. They commute because they act in the two different tensor components of the tensor product of two vector spaces. $R(0)$, the scalar Riemann curvature is a scalar constant and it therefore commutes with everything. It should also be noted that $C^T = -C$ where the transpose is taken w.r.t the first two indices of the Riemann curvature tensor which is antisymmetric w.r.t the exchange of these two indices. Now calculating the supertrace $\int str(K_t(x,x))dx$ of the kernel $K_t(x,y)$ of $exp(-tX)$, obtained by replacing the origin $x = 0$ by general x would give us the index of the Dirac operator for the following reason. Calculating the supertrace $str(K_t(x,x))$ involves just identifying the coefficient of $e^1 \wedge ... \wedge e^n$ in it, as observed earlier, where n is the dimension of the Riemannian manifold $\mathcal{M}$. The Dirac operator can be expressed as

$$D = \begin{pmatrix} 0 & D_- \\ D_+ & 0 \end{pmatrix}$$

where $D_- = D_+^*$ and hence

$$D^2 = \begin{pmatrix} D_-D_+ & 0 \\ 0 & D_+D_- \end{pmatrix}$$

So if we define the index of the Dirac operator as $ind(D_+)$, then we have

$$ind(D_+) = n(D_+) - n(D_-) = n(D_-D_+) - n(D_+D_-)$$
$$= Tr(exp(-tD_-D_+)) - Tr(exp(-tD_+D_-)) = str(exp(-tX))$$

We are now interested in computing

$$exp(-t(\partial^T\eta\partial - (1/2)x^TC\eta Cx - (3/2)x^TC\eta\partial + R(0) + K))$$

then computing its supertrace, then replacing C by $R_{\mu\nu}(x)$ and integrating over x. Finally we may set $t = 1$. That would result in the index of D_+. Since C is real and skewsymmetric, it can be by choosing an appropriate basis, brought into a direct sum of 2×2 matrices of the form

$$\begin{pmatrix} 0 & b \\ -b & 0 \end{pmatrix}$$

We may rename $\eta^{1/2}C\eta^{1/2}$ as C still retaining its skew-symmetry and simultaneously rename x by $\eta^{-1/2}x$ so that ∂ becomes $\eta^{1/2}\partial$. Then the above expression becomes

$$exp(-t(\partial^T\partial - (1/2)x^TC^2x - (3/2)x^TC\partial)).exp(-tK).exp(-tR(0))$$

Note that K commutes with C and hence also with C^2. We also observe that in the above 2×2 decomposition of C, C^2 has the decomposition into 2×2 matrices of the form $diag[-b^2, -b^2]$. Let $b = ia$. Computing the kernel of the above exponential then amounts to computing the kernel of the operator

$$exp(t(-\partial_x^2 - \partial_y^2 + (1/2)a^2(x^2 + y^2) + (3at/2)(x\partial_y - y\partial_x)))$$

Now, in polar coordinates,

$$x = r.cos(\phi), y = r.sin(\phi), x\partial_y - y\partial_x = \partial/\partial\phi$$

and $\partial/\partial\phi$ commutes with $(-\partial_x^2 - \partial_y^2) + (1/2)a^2(x^2 + y^2)$ and has eigenfunctions $exp(im\phi)$ with eigenvalues im. The trace of $exp(\lambda\partial/\partial\phi)$ is the same as the trace of a rotation of a function $f(\phi)$ by an angle $lambda$ giving the function $f(\phi + \lambda)$. Obviously this operator has eigenvalues $exp(i\lambda m), m \in \mathbb{Z}$ and

$$\sum_{m=-M}^{M} exp(i\lambda m) = sin(\lambda(M + 1/2))/sin(\lambda/2)$$

and

$$\int f(\lambda)sin(\lambda x)d\lambda/sin(\lambda)$$

$$= \int f(\lambda/x)sin(\lambda)d\lambda/x.sin(\lambda/x)$$

which converges as $x \to \infty$ to $\pi f(0)$. So $sin(\lambda(M + 1/2))/sin(\lambda/2)$ converges as $M \to \infty$ to $\pi\delta(\lambda)$. Unfortunately, this argument cannot be used here since $\lambda = 3a/2$ is a pure imaginary number. To make sense of this, we again use a scaling argument causing the angle of rotation ϕ to converge to zero and hence the operator $exp(\lambda\partial/\partial\phi)$ becomes in this limit the identity operator. Thus, the supertrace of the above kernel amounts to calculating first its ordinary trace treating a as an imaginary number and then replacing a by the supersymmetric matrix iR and then calculating the supertrace by extracting out the coefficient of $e^1 \wedge ... \wedge e^n$ in the resulting expression. The trace of the relevant kernel is then

$$Tr(exp(t(-\partial_x^2 - \partial_y^2 + (a^2/2)(x^2 + y^2))).Tr(exp(-tK))$$

and the supertrace then becomes replacing the product with $Tr(exp(-tK))$ with a wedge product and integrating over $\mathcal{M}$ which then automatically picks up the coefficient of $e^1 \wedge .. \wedge e^n$ resulting in the Aityah-Singer index theorem. Also it is clear that

$$Tr(exp(t(-\partial_x^2 - \partial_y^2 + (a^2/2)(x^2 + y^2))) =$$

$$[Tr(exp(t(-\partial_x^2 + a^2x^2/2))]^2$$

The exponential here is evidently the evolution kernel of the one dimensional quantum harmonic oscillator. Writing this Hamiltonian as

$$H = -\partial_x^2 + a^2x^2/2$$

with the transformation of variables $x = ky$ gives us

$$H = -k^{-2}\partial_y^2 + a^2k^2y^2/2 = 2k^{-2}((-1/2)\partial_y^2 + (a^2k^4/4)y^2)$$

We choose

$$k = 2^{1/4}\sqrt{(1/a)}$$

and then get

$$H = 2k^{-2}H_0 = a\sqrt{2}.H_0$$

where

$$H_0 = (-1/2)\partial_y^2 + y^2/2$$

This is the Hamiltonian of the one dimensional quantum harmonic oscillator in standard form. We define the oscillator creation and annihilation operators

$$a = 2^{-1/2}(y + \partial/\partial y), a^* = 2^{-1/2}(y - \partial/\partial y)$$

and find that

$$H_0 = a^*a + 1/2 = aa^* - 1/2$$

and by standard quantum harmonic oscillator analysis,

$$|n >= a^{*n}|0 > /\sqrt{n!}, n = 0, 1, 2, ...$$

$$a|0 >= 0, ie, |0 >= C.exp(-y^2/2)$$

For

$$< 0|0 >= 1,$$

(w.r.t the measure dy), we require

$$C = \pi^{-1/2}$$

$$a^{*n}|0>=\pi^{-1/2}2^{-n/2}(y-\partial/\partial y)^n exp(-y^2/2)$$
$$=2^{-n/2}.\pi^{-1/2}(-1)^n[exp(y^2/2)(\partial/\partial y).exp(-y^2/2)]^n.exp(-y^2/2)$$
$$=exp(-y^2/2).2^{-n/2}\pi^{-1/2}H_n(y)$$

where

$$H_n(y)=(-1)^n exp(y^2)(\partial^n/\partial y^n).exp(-y^2)$$

are the Hermite polynomials. Translating these formulae to the x-variable after incorporating the appropriate change of measure so that the new kets are normalized w.r.t. the measure dx gives us

$$|n>=2^{-n/2}.2^{1/4}(a\pi)^{-1/2}.H_n(2^{-1/4}\sqrt{a}.x).exp(-ax^2/\sqrt{2})$$

We thus evaluate in the y-domain,

$$exp(-tH)=exp(-ta\sqrt{2}H_0)$$

and

$$exp(-tH_0)(y,y')=\pi^{-1}\sum_{n\geq 0}exp(-(n+1/2)t)2^{-n}H_n(y)H_n(y')/n!$$

Now,

$$\sum_n z^n H_n(y)/n!=exp(y^2)\sum_n((-z)^n/n!)\partial_y^n exp(-y^2)$$
$$=exp(y^2-(y-z)^2)=exp(z^2-2yz)$$

Then,

$$\sum_n s^n z^n H_n(y)/n!=exp(2ysz-s^2z^2)$$

so that using

$$\int z^n(\bar{z})^m exp(-|z|^2)d^2z\int r^{n+m}exp(i(n-m)\phi)exp(-r^2)rdrd\phi$$

$$=\delta[n-m].2\pi.\int_0^\infty r^{2n+1}.exp(-r^2)dr=\delta[n-m].2\pi.\int_0^\infty x^n.exp(-x)dx/2$$
$$=\pi.n!.\delta[n-m]$$

Thus, we get

$$s^m H_m(y)=(1/\pi)\int exp(-s^2z^2+2syz-|z|^2)\bar{z}^m d^2z$$

so that

$$I(s,y,y')=\sum_m s^{2m}H_m(y)H_m(y')/n!$$
$$=(1/\pi)\int exp(-s^2z^2+2syz-|z|^2).exp(-\bar{s}^2z^2+2y's\bar{z})d^2z$$

The exponent of the integrand above is evaluated using $z = u + iv$ as

$$-s^2z^2 + 2syz - |z|^2 - s^2\bar{z}^2 + 2y's\bar{z} =$$

$$= -2s^2(u^2 - v^2) - (u^2 + v^2) + 2s(y + y')u + 2is(y - y')v$$

$$= -(1 + 2s^2)u^2 - (1 - 2s^2)v^2 + 2su(y + y') + 2isv(y - y')$$

and hence

$$\pi.I(s/\sqrt{2}, y, y') = \int exp(-(1+s^2)u^2-(1-s^2)v^2+\sqrt{2}su(y+y')+i\sqrt{2}sv(y-y'))dudv$$

$$= 2\pi(2(1+s^2))^{-1/2}(2(1-s^2))^{-1/2}exp(s^2(y+y')^2/2(1+s^2)-s^2(y-y')^2/2(1-s^2))$$

Now,

$$s^2(y+y')^2/(1+s^2) - s^2(y-y')^2/(1-s^2) = (y^2+y'^2)(s^2/(1+s^2) - s^2/(1-s^2))$$

$$+2yy'(s^2/(1+s^2) + s^2/(1-s^2))$$

$$= (-2s^4/(1-s^4))(y^2+y'^2) + 2yy's^2/(1-s^4)$$

Thus, we get

$$I(\sqrt{s/2}, y, y') = \sum_m (s^m/2^{m/2})H_m(y)H_m(y') =$$

$$(1-s^2)^{-1/2}.exp((-s^2/(1-s^2))(y^2+y'^2) + yy's/(1-s^2))$$

and then, with $s = exp(-t)$,

$$exp(-tH_0)(y, y') = (1-s^2)^{-1/2}\sqrt{s}exp(-(y^2+y'^2)/2)\sum_n 2^{-n/2}s^nH_n(y)H_n(y')/n!$$

$$= (1-exp(-2t))^{-1/2}.exp(-t/2).exp((-(1+s^2)/2(1-s^2))(y^2+y'^2)+yy's/(1-s^2))$$

We observe that

$$(1+s^2)/(1-s^2) = (1+exp(-2t))/(1-exp(-2t)) = coth(t),$$

$$s/(1-s^2) = exp(-t)/(1-exp(-2t)) = 1/(exp(t) - exp(-t)) = 1/2.sinh(t)$$

So,

$$exp(-tH_0)(y, y') = \sqrt{2}(sinh(t))^{-1/2}.exp(-coth(t)(y^2+y'^2) + yy'/2.sinh(t))$$

In the $x - x'$ coordinate system, we then find that after taking into account the change of measure $dx = 2^{1/4}\sqrt{(1/a)}dy$

$$exp(-tH)(x, x')|_{x=x'=0} = exp(-ta\sqrt{2}H_0)(x, x') = (a\sqrt{2})^{1/2}(sinh(ta\sqrt{2}))^{-1/2}$$

Since the supertrace is invariant under scaling, we are setting $x = x' = 0$. Thus, evaluating this at $t = 1/\sqrt{2}$, we find that the Dirac index is given by the integral of

$$det(sinh(iC)/iC)^{-1/2}$$

where C is to be taken as $R(x) = R_{\mu\nu}(x)e^{\mu} \wedge e^{\nu}$. Thus, the formula for the index of the Dirac operator becomes

$$ind(D_+) = \int_{\mathcal{M}} [det(sinh(iR(x))/iR(x))]^{-1/2} \wedge Tr(exp(-K(x)))$$

with

$$K(x) = K_{\mu\nu}(x)e^{\mu} \wedge e^{\nu}$$

and

$$Tr(exp(-K(x))) = \sum (r!)^{-1} Tr(K_{\mu_1\nu_1}(x)K_{\mu_2\nu_2}(x)...K_{\mu_r\nu_r}(x))) e^{\mu_1} \wedge e^{\nu_1} \wedge ... \wedge e^{\mu_r} \wedge e^{\nu_r}$$

Remarks:To calculate directly the trace of the kernel, $exp(-coth(t)(y^2 + y'^2) + yy'/2.sinh(t))$ without making use of scaling, we would set $y' = y$ and integrate over y. That would give us

$$\int exp(-y^2(2coth(t) + 1/2.sinh(t)))dy$$

and it would evaluate to something proportional to $(2.coth(t)+1/2.sinh(t))^{-1/2}$. We have to replace t by $itC\sqrt{2}$ or equivalently by itR where R is a two form giving $(2.coth(itR) + 1/2.sinh(itR))^{-1/2}$. This is the the square root of an odd function of the two form R and does not make much sense. However by using the scaling argument involving replacement of y by zero, we just get unity and the supertrace is then calculated by replacing the centre zero of localization by an arbitary x and then evaluating the supertrace as an integral over x with products in the integrand being taken using the exterior product from which the coefficient of $e^1 \wedge ... \wedge e^n$ has been extracted.

Interpretation the Atiyah-Singer index theorem in terms of gauge anomalies in non-Abelian quantum field theory.

Consider a path integral

$$\int exp(iI(\psi))D\psi D\psi^*$$

the action $I(\psi)$ is gauge invariant but the path measure $D\psi.D\psi^*$ is not gauge invariant. In fact, a gauge transformation corresponds to multiplying ψ by $exp(i\epsilon\gamma_5 \times t)$ if we assume that t is a generator of the the non-Abelian gauge group and $\gamma_5 = \gamma^0\gamma^1\gamma^2\gamma^3$ is the Chiral factor. For small ϵ, it is clear that the path measure $D\psi.D\psi^*$ gets altered by the trace of the kernel $\epsilon\gamma_5 \times t\delta(x - y)$ mutiplied by the path measure. This is because $det(exp(\epsilon\Lambda)) = exp(\epsilon.Tr(\Lambda)) = 1 + \epsilon.Tr(\Lambda) + O(\epsilon^2)$. Now the trace of the kernel $\delta(x - y)$ is $\int \delta(x - x)dx$ which does not make sense. So we replace $Tr(\delta(x - y))$ by $M^2.Tr(f((\gamma.D_x)^2/M))$

where D_x is the gauge covariant derivative $\gamma.(\partial + ieA)$ with A the non-Abelian gauge potential. In the limit as $M \to \infty$, this becomes $Tr(f(0)I)$which corresponds exactly to what we want. The additional feature of making such an approximation to the trace of the δ function is that that it is a gauge invariant expression since its is built out of the gauge covariant derivative. The only term that contributes to the above trace is $(1/2)f''(0)Tr((\gamma.D_x)^4))$. Now,

$$D_x^2 = (\gamma.(\partial + ieA))^2 =$$

$$\gamma^\mu\gamma^\nu D_\mu D_\nu = \eta^{\mu\nu}D_\mu D_\nu + (1/2)[\gamma^\mu, \gamma^\nu]F_{\mu\nu}$$

where

$$F_{\mu\nu} = [D_\mu, D_\nu]$$

Then,

$$Tr(D_x^4) = \int [(1/4)Tr([\gamma^\mu, \gamma^\nu][\gamma^\rho, \gamma^\sigma])Tr(F_{\mu\nu}F_{\rho\sigma})]d^4x$$

$$= \int \epsilon(\mu\nu\rho\sigma)F^a_{\mu\nu}F^a_{\rho\sigma}d^4x$$

This quantity therefore measures the anomaly, the amount of path measure that is not conserved. It is also related to violation of current conservation. In terms of the Yang-Mills curvature tensor

$$F^a = F^a_{\mu\nu}dx^\mu \wedge dx^\nu$$

we can also write this as

$$\int F^a \wedge F^a$$

and by the Atiyah-Singer index theorem for flat space-time with $K = F^a t_a$, we have $Tr(K \wedge K) = F^a \wedge F^b Tr(t_a t_b) = F^a \wedge F^a$ since $Tr(t_a t_b) = \delta_{ab}$. and the above integral is the index of the Dirac operator. Another way to express this is as follows: Consider the eigendecomposition of $\gamma.D$):

$$\gamma.D = \sum_k c(k)\phi_k\phi_k^*$$

Since γ^5 anticommutes with $\gamma^\mu, \mu = 0, 1, 2, 3$, it follows that γ^5 also anticommutes with $\gamma.D$. Hence, if ϕ_k is an eigenvector of $\gamma.D$ with eigenvalue $c(k)$, then $\gamma_5\phi_k$ is also an eigenvector of $\gamma.D$ but with eigenvalue $-c(k)$. We can then write the above spectral decomposition as

$$\gamma.D = \sum_{k:c(k)\in E} c(k)(\phi_k.\phi_k^* - (\gamma^5\phi_k).(\gamma^5\phi_k)^*)$$

Let ϕ be an eigenvector of $\gamma.D$ corresponding to the zero eigenvalue. Then $\gamma^5\phi$ is also an eigenvector of $\gamma.D$ corresponding to the zero eigenvalue. Since $(\gamma^5)^2 = I$, $(1+\gamma^5)\phi$ and $(1-\gamma^5)\phi$ are eigenvectors of γ^5 belonging to eigenvalues 1 and -1 respectively. Let n_+ denote the number of eigenvectors of $\gamma.D$ having zero eigenvalue but which are eigenvectors of γ^5 belonging to the eigenvalue 1 and n_- the number of eigenvectors of $\gamma.D$ having zero eigenvalue but which are eigenvectors of γ^5 belonging to the eigenvalue -1. Then, obviously

$$Tr(\gamma^5 I) = n_+ - n_-$$

where $\gamma^5 I$ acts as γ^5 on the space spanned by the eigenvectors of $\gamma.D$. Thus, we have proved a special case of the Atiyah-Singer index theorem corresponding to flat space-time:

$$n_+ - n_- = \int \epsilon(\mu\nu\rho\sigma) F^a_{\mu\nu} F^a_{\rho\sigma} d^4x$$

Remark: $(1+\gamma^5)\phi_k$ has eigenvalue $+1$ for γ^5, and $(1-\gamma^5)\phi_k$ has eigenvalue -1 for γ^5. If $c(k) \neq 0$, then $(1+\gamma^5)\phi_k$ and $(1-\gamma^5)\phi_k$ are both non-vanishing eigenvectors of γ^5 with eigenvalues ± 1 respectively and hence their contribution to the trace of $\gamma^5 I$ vanishes. To see that they are both non-vanishing, we just observe that $(1-\gamma^5)\phi_k = 0$ implies

$$0 = \gamma.D(1-\gamma^5)\phi_k = (1+\gamma^5)\gamma.D = c(k)(1+\gamma^5)\phi_k$$

which implies that $(1+\gamma^5)\phi_k = 0$ and hence $\phi_k = 0$, a contradiction. Thus, contribution to $Tr(\gamma^5 I)$ comes only from the zero modes of $\gamma.D$.

Appendix

Note: By use of the Bianchi identity for the Riemann curvature tensor, we find that

$$R_{abcd} + R_{acdb} + R_{adbc} = 0, R_{cdab} = R_{abcd}, R_{bacd} = -R_{abcd} = R_{abdc}$$

and hence

$$R_{abcd}\gamma^c\gamma^d\gamma^a\gamma^b = R_{abcd}\gamma^a\gamma^b\gamma^c\gamma^d =$$
$$= -R_{abcd}\gamma^a(\gamma^c\gamma^d\gamma^b + \gamma^d\gamma^b\gamma^b\gamma^c)$$

and then by using the property

$$\gamma^a\gamma^b = 2\eta^{ab} - \gamma^b\gamma^a$$

of the Dirac matrices, it is easily deduced from the above identity that

$$R_{abcd}\gamma^c\gamma^d\gamma^a\gamma^b = 2R$$

where

$$R = \eta^{bc}\eta^{ad}R_{abcd}$$

is the scalar Riemann curvature.

Note:Covariant differentiation of vector fields in the spinor representation is defined as follows:

$$[\nabla_\mu, \gamma^\nu B_\nu] = [\partial_\nu + \Gamma_\nu, \gamma^a B_a]$$
$$= \gamma^a B_{a,\nu} + [\Gamma_\nu, \gamma^a] B_a$$
$$[\Gamma_\nu, \gamma^a] = [J^{bc} e_{b\nu:\mu} e^\nu_c, \gamma^a]$$

$$= [J^{bc}, \gamma^a] e_{b\nu:\mu} e_c^\nu$$

$$= (\eta^{ca}\gamma^b - \eta^{ba}\gamma^c) e_{b\nu:\mu} e_c^\nu$$

Thus,

$$[\Gamma_\nu, \gamma^a] B_a = [\gamma^b B^c - \gamma^c B^b] e_{b\nu:\mu} e_c^\nu$$

$$= [\gamma^b B^\nu - \gamma^\nu B^b] e_{b\nu:\mu}$$

$$= (\gamma^\rho B^\nu - \gamma^\nu B^\rho) e_\rho^b e_{b\nu:\mu}$$

$$= \gamma^\rho B^\nu (e_\rho^b e_{b\nu:\mu} - e_\nu^b e_{b\rho:\mu})$$

$$= \gamma^\rho B^\nu (e_\rho^b e_{b\nu,\mu} - e_\nu^b e_{b\rho,\mu})$$

$$+\gamma^\rho B^\nu (e_\nu^b \Gamma_{\rho\mu}^\alpha e_{b\alpha} - e_\rho^b \Gamma_{\nu\mu}^\alpha e_{b\alpha})$$

$$= \gamma^\rho B^\nu (e_\rho^b e_{b\nu,\mu} - e_\nu^b e_{b\rho,\mu})$$

$$+\gamma^\rho B^\nu (g_{\nu\alpha} \Gamma_{\rho\mu}^\alpha - g_{\rho\alpha} \Gamma_{\nu\mu}^\alpha)$$

$$= \gamma^\rho B^\nu (e_\rho^b e_{b\nu,\mu} - e_\nu^b e_{b\rho,\mu})$$

$$+\gamma^\rho B^\nu (\Gamma_{\nu\rho\mu} - \Gamma_{\rho\nu\mu})$$

References

[1] Steven Weinberg, "The quantum theory of fields, vol.III, Supersymmetry", Cambridge University Press.

[2] V.S.Varadarajan, "Supersymmetry for Mathematicians, An Introduction", Courant Institute Lecture notes.

[3] M.Green, J.Schwarz and E.Witten, "Superstring Theory", Vols. I and II, Cambridge University Press.

[4] K.R.Parthasarathy, "An introduction to quantum stochastic calculus", Birkhauser.

[5] Timothy Eyre, "Quantum Stochastic Calculus and Representations of Lie Superalgebras", Springer Lecture notes in Mathematics.